The Truth About Charlie

the "madman" stranded on Howick Island with Ion Idriess

Rob Coutts

First published 2013

National Library of Australia Cataloguing-in-Publication entry
Author: Coutts, Rob.
Title: The truth about Charlie / Rob Coutts.
ISBN: 9781922109767 (paperback)
Subjects: Idriess, Ion L. (Ion Llewellyn), 1889-1979.
Tritton, George., 1874-1928.
Frontier and pioneer life--Queensland--Howick Island.
Prospecting--Queensland--Howick Island.
Howick Island (Qld.)--History.
Dewey Number: 994.38

Typeset in Garamond 11 pt.
Cover Design: Boolarong Press

Published by Boolarong Press, Salisbury, Brisbane, Australia.
Printed and bound by Watson Ferguson & Company, Salisbury, Brisbane, Australia.

Dedication

For Bev.
The love of my life and my wife for fifty-two years.

And in memory of our son Greg.
Greg died at just forty-seven years of age in 2011.

Contents

Naturally, the first acknowledgement must be to my wife Bev who endured long hours by herself while this book was being written and rewritten. Bev also read the first draft and provided me with the first edit of the manuscript.

Then I must extend my unstinting thanks to Jim McJannett of Cooktown. Jim's expertise as an historical researcher mainly centres on the life and times of John Dickie, the legendary North Queensland bushman/prospector. However, in the course of decades of research Jim has spoken with people who knew George Tritton. Extraordinarily, Jim spoke with Jack Idriess and asked him about the Howick adventure. Through innumerable emails Jim provided much information about George Tritton and in particular, he raised the debate about the *Cairns Post* article explored in Chapter Thirty-seven. Jim will not agree with my conclusion but my book would have been poorer without his input.

I appreciated the help of long-time Cooktown residents Helen and Peter Rutherford for their knowledge of Howick Island. Helen and Peter have walked the island with the second *Madman's Island* book in hand. Among other features, they found the well and the rock upon which George irrigated his bowel. With reservations, Helen and Peter say the basic story-line of the second *Madman's Island* is true. Helen and Peter did all they could to get me on to the island but circumstances prevented the difficult journey. One day?

Acknowledged is the assistance of Beverley and John Shay of the Cooktown Historical Society for their information about Cooktown and for their permission to reprint the photograph of the West Coast Hotel (Preface). I also want to thank Wilfred (Willie) Gordon for his information about the Aboriginal people of the Cooktown area who visited Howick Island long before George and Jack were stranded there.

The Facts

As with any good story, some readers will question the accuracy of some of the detail. Certainly there is room for different interpretations. Other readers, however, will dispute even essential aspects of the story so it is important to set out the facts of the story. The first, indisputable, fact is that in 1920 two prospectors Ion Llewellyn (Jack) Idriess and Thomas George (George) Tritton travelled to a remote North Queensland island looking for tin.[1]

TWO – George and Jack were two bushmen and both were ex-servicemen of the First World War.

THREE – The island was Howick Island; an inhospitable speck of land surrounded by a mangrove forest 100 kilometres north-east of Cooktown.[2]

FOUR – George and Jack were stranded when the boat that delivered them to the island failed to return.[3]

FIVE – They lived on the island from September 1920 to when Jack was rescued. Jack was taken back to Cooktown in February 1921 but George continued living on the island for a short time: perhaps a few months.[4]

SIX – What actually happened during the stranding is not without controversy. George's side of the story is not known but Jack published his version in two books both titled *Madman's Island*.[5]

SEVEN – Although the basic story-line is similar in each book, the first book was written as fiction but Jack explicitly stated the second *Madman's Island* was written "*strictly according to fact*".[6]

EIGHT – However, a newspaper report of Jack's statement to the police on his return to Cooktown differs significantly from the story he told in his books.[7] This newspaper article is discussed in detail in Chapter Thirty-seven.

NINE – Despite Jack's assertion that the second *Madman's Island* book was factual, both of the *Madman's Island* books contain material added for literary effect.

TEN – Even though the basic story-line is similar in each of the *Madman's Island* books, Jack Idriess was a master story-teller. He eventually published fifty-three books and innumerable press and magazine articles. He always had an eye for a good story.

Mark Twain eloquently explained the practice of embellishing a story for literary effect. In *The Adventures of Huckleberry Finn*, Mark Twain had Huck saying:

> "*You don't know me, without you have read a book by the name of The Adventures of Tom Sawyer, but that ain't no matter. That book was made by Mr. Mark Twain, and he told the truth mainly.* <u>*There were things which he stretched, but mainly he told the truth*</u> (underlining added). *That is nothing. I never seen anybody but lied, one time or another...* (but Mark Twain's book)... *is mostly a true book; with stretchers, as I said before*".

Borrowing Mark Twain's words, there were parts of the story of the stranding on Howick Island that Jack stretched, but the basic story is fact.

ELEVEN – In his first, 1927, book Jack did not tell his readers the real name of the island but in the second, 1938, *Madman's Island* he did declare it as Howick Island. In both books Jack called George Tritton, "Charlie".

TWELVE – George (as "Charlie") was portrayed by Jack as the madman of Madman's Island.

THIRTEEN – Facts about George's life are scarce but there is more information available about Jack. A biography, *Ion Idriess*, was written by Beverley Eley.[10]

FOURTEEN – Eley's book has been criticised by readers interested in Jack's life and there have been a number of literary reviews of the book including one that was not at all favourable.[11] Even so, for anyone wanting to know about Jack's life Eley's book is an essential starting point, not least

because there is little to be found about Jack's life and work from other sources.[12] Because of her privileged access to Jack's diaries and to members of Jack's family, the Eley biography is important to try to get some insight into the personality of the man who fought with "Charlie" on Howick Island.

Outside these few facts, what actually happened on Howick Island in 1920/1921 will never be known – both men are long dead – so this book examines the story as Jack told it.

Preface

Let him die! That's what it came down to on "Madman's Island". On the tiny and inhospitable Howick Island, Jack Idriess and George Tritton had been fighting each other for months. Jack was sure George was about to die a horrible death sinking in liquid mud but he just watched the struggle without trying to help.

"*Curse him; why should I help him anyway*?" Jack said to himself.[1]

It is hard to imagine how it could to the point where one man could just look on as another man struggles for life. After four months stranded on a tiny scrap of land, Jack watched the desperate plight of the only other person on the island and said to himself, "*Sooner or later he or I must die. It's survival of the fittest.*"[2] Of course the background to Jack's inactivity was complex: the relationship between the two men had been steadily worsening for months.

But there it was – let him die!

The battlefields of the First World War were part of the background to the story. Jack and George were both veterans of the First World War and both had suffered severe injuries. Jack was wounded three times.[3] Although George's condition was not a war-wound,[4] his ongoing disability was much more severe – he had a substantial part of his bowel removed and faeces had to be flushed through a hole in his abdomen. Another part of the background was that both men were experienced and resourceful bushmen. Both had prospected the wild jungles of the Cape York Peninsula in far North Queensland.

So, when their relationship soured, two men of the bush who had actual battlefield experience were pitted one against the other on Howick Island. Even though George thought he might live on the island permanently the island is not really liveable in a practical sense.

The two men nearly died there.

Their adventure started with Jack and his mate Dick Welsh having a beer (or three) at their favourite pub – Cooktown's West Coast Hotel pictured here in 1920: the very year in which Jack and George decided to go prospecting for tin.

The West Coast Hotel was built on its present site around 1874; just a year after Cooktown was born. It has survived the wild fluctuations of Cooktown's economic fortunes (and even wilder tropical cyclones) to continue trading today. In the bar Jack and Dick Welsh met George who was serving behind the bar. George produced some black stones and the three prospectors agreed the stones were a mixture of tin and wolfram. Jack wrote that the stones had been found on Howick Island by a Malay fisherman they called "Old Tarquay".

The conversation in the West Coast Hotel probably occurred around July in 1920, a couple of months before Jack's thirty-first birthday and George was forty-six. Dick Welsh decided not to go but Jack and George were left on Howick Island in September 1920. They were stranded when the boat that delivered them to the island did not return.

For a couple of weeks everything was calm but inevitably, there was conflict between them. The two men lived together for months on a tiny island. Before they arrived on Howick Island they did know each other but they were not regular prospecting mates. Both of them (in their different ways) were self-contained, introspective loners. In retrospect, the trip was ill-conceived but it was too late – they were stranded as their relationship worsened until they fought. George tried to shoot Jack. Jack later tried to kill George.

Did George go "mad" as Jack claimed or was George unfairly blamed for the conflict between the two men? In the answers to these questions is the Truth about Charlie.

Part One

George and Jack and Howick Island

1

George's early life

Although George was born in Cork, Ireland, in 1874, both his parents were English and he grew up in Devon.[1] Devon is bordered in the west by Cornwall and in the east by Dorset and Somerset. Devon's southern coast is on the English Channel and the northern coast (where George's mother and father were married) is across the Bristol Channel from Ireland where George was born. At 111 kilometres in length and ninety-six kilometres wide, Devon is the fourth largest English county.

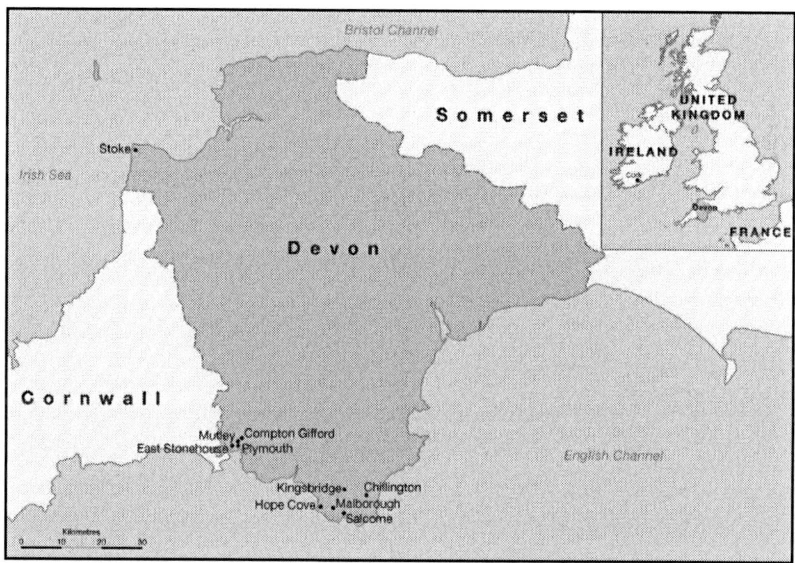

George grew up in a tiny fishing town called Hope Cove where his father, William Tritton, was one of the Coastguards. William Tritton was born and grew up in Woolwich, not far from London.[2] His mother (Elizabeth Scott before she married) lived almost all of her life in Devon except for a few years in Ireland where her husband had moved for his work. Even when (much later in Australia) George joined up for service in the First World War, Elizabeth was still living in Devon.[3]

In his early twenties George's father, William, moved to Devon where he met his future wife. George's parents were married in Stoke in 1869.[4] Over the course of nineteen years Elizabeth Tritton bore ten children – six boys and four girls.[5]

George's older brothers were Washington, born in Kent (near where William Tritton was born) and William who was born in East Stonehouse (where Elizabeth Tritton was born). As it happens, it was these two brothers with whom George (much later) travelled to Queensland.

Then in 1874 George was born at Cork in Ireland. William Tritton was working there as a coastguard. The family spent only three years in Ireland but during that period two other children (Albert and Jenette) were also born. After their stay in Ireland the growing Tritton family moved to the Hope Cove Coastguard Station on the coast of Devon near Malborough. The next two children, Theopilos and Elizabeth, were both born at Hope Cove.[6]

Hope Cove is really two villages – Inner Hope and Outer Hope – connected by a cliff-top path. When George was a child the only road to the outside world was the winding cart track to Galmpton.[7] When the Tritton family lived at Inner Hope, the little town had about thirty houses and 150 people while in Outer Hope there were about twenty houses and sixty-four people.[8] George and his family lived in a Coastguard cottage at Inner Hope.[9]

In those days Hope Cove was a wild, isolated place where people had to be tough, self-sufficient and above all, resilient. Along that wild coast in the 1880s life was hard. There was no electricity, just candles and hurricane lamps. Water had to be drawn from a village pump and carried home in a bucket. Supplies for the village came by horse and cart from Salcombe, about eleven kilometres away or about nine kilometres from Kingsbridge.[10]

With Washington and Bill, as the eldest children, George was progressively expected to help support the family. The children helped their father grow

vegetables and keep chickens that had to be fed by the light of a hurricane lamp before the children went to school[11] at the Church of England School at Malborough.[12] Some children also went stone-picking for twopence a day. Stone-picking was laboriously picking up stones and piling them into heaps to clear fields for cultivation. All the village children would regularly check the beach for driftwood. Any piece of wood over 2.5 metres in length had to be declared to the Receiver of Wrecks but the finder could get one-third of its value.[13]

The main (legitimate) industry of Hope Cove was fishing – mainly for crabs and lobsters but in August each year pilchard shoals arrived in the bay. When the shoals were spotted, the large flat-bottomed vessels with oars and a sail were hurriedly launched to draw a net in a circle around the shoal. The catch was raced off to markets at Salcombe or Kingsbridge in small fish-carts pulled by Dartmoor ponies.[14] In those days the whole village relied on fishing; seeking pilchards, mackerel, turbot, lobsters and crabs[15] but it was not always so. Smuggling was also lucrative.

To counter the illegal trade, the Coastguard service was established under the Board of Customs. Terraces of houses were built at places around the coast where smuggling was most likely. One of those places was Hope Cove. The first row of terraces was at Outer Hope but by the time the Tritton family shifted into one of these cottages in 1877, the Coastguard operations had moved to Inner Hope. William Tritton became one of the eight Coastguards serving under a Royal Navy Lieutenant.[16]

The Coastguards' standing in the community was ambivalent. On the one hand they were directly responsible for preventing the smuggling of contraband: an action that adversely affected the local "black" economy. At the same time, with Hope Cove's volunteer lifeboat men, the Coastguards were responsible for helping save lives at sea.[17]

Many Hope Cove families supplemented their fishing income by helping (directly or indirectly) in the smuggling enterprise. Three-masted luggers sailing from France with tobacco, silks and liquor (particularly brandy) would stand offshore before lowering kegs to be transported to shore when the Coastguards were not around.[18] For a major shipment just about everyone in the village would have a job. The smugglers needed people to man the

small boats and people to carry the cargo when it was ashore. They might sometimes have needed horses to carry it. And, of course, the smuggled cargo had to be hidden until it could be sold. Some of the little thatched-roof stone cottages even had special windows from which a light would warn smugglers if the government men were around.

Open conflict between smugglers and the Coastguard was always likely. On 2 November 1840 the London *Times* reported a serious incident further east along the coast from Hope Cove. On that occasion local boats being used to bring smuggled goods ashore were seized by the Coastguard. This seizure was, of course, a big problem for the village and the villagers attacked the Coastguard crew. About forty men battled in the pitch-black darkness up and down the cliffs. The officer in charge nearly drowned and all of the Coastguard crew were badly beaten. The smugglers escaped. The *Times* newspaper had other reports of serious incidents of confrontation between smugglers and government officers.[19]

On another occasion eighteen Hope Cove smugglers' wives were caught by the Coastguard preparing donkeys to transport the contraband. They were arrested and sent to Exeter gaol for insulting the Coastguard men. The fishermen/smugglers escaped and sold the smuggled goods.[20]

Along this wild and isolated coastline stories of smuggling and shipwrecks were common but by the time George was born the difficult-to-get-to isolated town was starting to attract touring visitors. Nevertheless, it was still a hard and precarious life for those who lived there. As a child George would have known the children of smugglers. He grew up in a village where co-operation with smugglers was part of scraping a living from that harsh coast. The people faced constant conflict not only with the officers of the government but also with the stormy sea and rugged coastline of the area. Shipwrecks were common and indeed the area is still today regarded as hazardous to shipping.

The problem is that Bigbury Bay, a wide shallow curve of coastline in South Devon with Hope Cove at its easterly end, is a trap for unwary ships whenever there is a southerly wind. The area around Hope Cove then becomes a lee shore but the danger is not obvious until it is too late to steer out of the bay to safety. This wild coastline around Hope Cove was renowned for its shipwrecks and this was another side to William Tritton's work.

Danger for the coastguard crew was always present. Although William Tritton was not part of the particular crew that went to the aid of a ship which had run aground further along the coast, the incident illustrates his duties. On 31 January 1845 at Dungeness, Kent, the *William Harrington* went aground at low tide during a severe snow storm. The Coastguard crew of four men rowed out to the ship when the snow cleared but there was a strong wind and high sea. The tide was then running strongly. The rowers had a desperate struggle to get out to the ship. As the men got close to the ship and the officer in charge was climbing aboard, the Coastguard boat capsized. Two of the four men in the boat were drowned. Another Coastguard boat was launched. Eventually the ship was saved but only at the cost of the two lives of the men in the Coastguard crew.[21]

Fishermen who worked along this coast were in constant danger from the wild sea. In one incident in 1882, although the Coastguard men were quickly on the scene of a capsized fishing boat, two men had already drowned. The son of one of the men was clinging to the rigging of the capsized boat and was rescued.[22] In 1883, just a year after this spectacular rescue George's father, William Tritton was retired with a pension.[23] It is more than likely that he was injured on duty in some similar rescue.

At the time William was granted his pension universal age pensions were still being debated in the British Parliament.[24] So, although George's father was lucky that he was entitled to a pension, the family's income was much reduced. Their economic situation would have become even more desperate when they had to leave the Coastguard cottage at Outer Hope. Although life was never easy for the Tritton family, in 1883 things became a whole lot worse. In 1884 the Trittons shifted to Chillington eight miles east of Hope Cove where yet another child, Charles, was born.[25]

This, then, was the world of George's childhood. Until he was ten he lived on one of the most isolated and dangerous coasts in the world among hard men who were on the edge of lawlessness. Life was tough and so were the people. Bravery and violent death were accepted as a normal part of living. Then when his father was hurt he had to leave school to start helping support the large Tritton family.

By 1886 the Trittons were living in Mutley (near Plymouth) on William

Tritton's pension and there were nine children in the family ranging in age from Washington at sixteen years of age to the newly-born Florence. Something had to be done. There were too many of them.

For a long time the family had been hearing stories about Australia: about gold and horses with golden horseshoes.[26] They would have discussed Australia and specifically Queensland because the three older boys did in fact arrive in Townsville. Why not? One Englishman said Cooktown was the best-known place in all of Australia.[27] People from all over Australia, from California and from China flocked to Cooktown.[28]

So in 1888 at only fourteen years of age George and his two older brothers Washington (18) and William (16) set off for Queensland. The gold was real enough and although George and his brothers found later that many of the stories were nonsense, some of them were true. Anyway, true or not, the stories were reason enough to get them on their way. At the time they did not know they were too late. The Palmer River gold was gone even before they left home.

There is no record of how the boys got the money for their passage but in 1888, with Washington and William, George travelled to London on the first leg of the journey to the other side of the world. They boarded the *Quetta* in April 1888. In one sense they were lucky. Washington, William and George landed safely in Townsville on 28 May 1888[29] but only two years later the ill-fated *Quetta* became part of Australia's maritime history. On its very next trip, in February 1890, the *Quetta* was on the return trip to London when it was wrecked in the Torres Strait and sank within minutes. Of the 290 people on board, 133 died.[30]

George would have learned at some later date that in 1892 his father had died at only forty-nine years of age.[31] His mother remained in the Plymouth area. She bore her tenth child in 1889 – a year after George left England and only a few years before his father died.[32]

When George arrived in Australia, Jack was still more than a year away from being born (on 20 September of the following year). Indeed George and his brothers landed in Townsville about six months before Jack's parents were to be married on 1 December 1888.

2

Jack's early life

Fifteen years after George was born in Ireland, Jack was born on 20 September 1889 at Waverley in New South Wales. According to Jack's biographer, Beverley Eley, Jack's father's name was Walter Owen Idriess[1] but this is not correct. Jack's father's real name is not known. Nevertheless, in this book Jack's father is referred to as Walter Owen Idriess because this is the name under which he was married and under which lived his life in Australia.

Jack's father joined the Royal Navy as Walter Owen Idriess and on his marriage certificate he declared himself to be Walter Owen-Idriess.[2] For some unfathomable reason, Jack's father seems to have adopted an old Welsh name. Walter's real family name might have been Jones but neither this name (nor Idriess nor Owen-Idriess) responds to searches using Walter's birth details and the names of his parents as provided on Walter's death certificate.[3] Whatever his given name, Jack's father used the name Walter Owen Idriess and visited Sydney on several occasions before he met the girl who would become Jack's mother, Julia Edmonds. Another confusing issue is that on Jack's birth certificate the name "Windeyer" appears.[4] Beverley Eley said Jack's full name was Ion Llewellyn Windeyer-Idriess[5] but Jack's family had only a tenuous claim on that name.

"Windeyer" was the second given name of Jack's grandfather John Windeyer Edmonds. His father John Windeyer 1779-1835 was Jack's great-grandfather. He had a relationship with Eleanor Edmonds. Their child was Jack's grandfather John Windeyer Edmonds 1833-1905.[6] His daughter was Jack's mother.

His mother's registered birth name was Julia Edmonds.[7] Eley said Jack's mother's name was Juliette Windeyer-Edmonds and suggested a connection between Julia's family (the Edmonds) and a much more distinguished Sydney family (the Windeyers).[7] Beverley Eley said Julia was the niece twice removed of Charles Windeyer and second cousin to Sir William Charles Windeyer.[8]

In fact, Jack's mother was born on 12 November 1865 at Woolshed, Cunningham Plains, near Binalong in New South Wales. Her birth was registered in 1866 at Binalong as Edmonds. Her father declared himself as John W Edmonds and not Windeyer-Edmonds.[9] In the *Yass Courier* newspaper Julia's birth was announced as the daughter of Mr John Windeyer Edmonds.

Nevertheless, Jack's birth was registered as "Ion Windeyer" (not Ion Windeyer Idriess or Ion Llewellyn Idriess – the name Idriess is not mentioned).[10] Either his father made some sort of mistake or he continued his habit of adopting a name to suit himself. At some later date Jack was given the name he lived with all his life – Ion Llewellyn Idriess.

So, given this mysterious confusion over Jack's family name it is not surprising that there is doubt about the way the name is pronounced. Many people assume the pronunciation of Jack's family name was Idr-EE-ess but the Australian Dictionary of Biography prefers Idr-UH-ss. Around the Daintree River south of Cooktown he is known as EYE-dress. And there may be other pronunciations. Even Jack's descendants pronounce the name differently but the authority must be the man who held the name – Jack himself. In an oral history interview in 1974 Jack seems to pronounce his name as EE-dress[11] and a press article based on an interview Jack gave in 1950 specifically stated that his name was pronounced EE-dress.[12]

For all of his life Jack acknowledged the name of Ion Llewellyn Idriess. Even so, he was known to his mates in the bush as Jack. It is likely that he started calling himself Jack before he was eighteen. Certainly, when he finally broke away from the care of his maternal grandfather's second wife Kate Eldershaw in 1908, he fully adopted the name Jack.

Except in official documents (as, for instance, when he joined up for service in First World War) Jack did not seem to use his real name again until his first book *Madman's Island* was published. In the 1938 rewriting of that book Idriess explains why he called himself Jack; "*In these covers I am alluded to as "Jack". That was always my old bush name; few of my old friends of the far-out lands ever knew me as Ion*".[13] He gave a similar explanation in other books. In the Author's Note to *Back o' Cairns* Jack said he was teased at school because the name Ion was too different to the usual names of the day. He did not like it. He said he called himself Jack when he ran away to sea to escape from the care of his step-grandmother. He said he kept the name Jack at his second (successful) attempt to run away.

So it appears Jack suffered the name Ion until he was eighteen then he changed his name when he made the break away from his family for a life in the bush. At another much later time he wrote of his dislike for the name Ion in a newspaper column. In this column, Jack's friend Alec Chisholm wrote, "*Ion Idriess respects the Ion because it has descended to him from his Welch forefathers but for everyday use he much prefers plain Jack*". These comments "*evoked heartfelt support*" from Jack who responded: "*What torment some of us would have been saved at the school roll call if our parents had had a nicer sense of the fitness of things.*"[14]

As late as 1958 Jack said he would still occasionally hear a disbelieving bushie growl, "*Jack Idriess! He didn't write them books! I knoo Jack Idriess well. He could no more write books than I can. Ion Idriess wrote them books – some city bloke!*"[15]

So Ion Llewellyn Idriess was "Jack" in the bush and "Ion" as an author – all his books are written under the name Ion Idriess or Ion L Idriess and only the Author's Note to *Back o' Cairns* is signed Ion ("Jack") Idriess.

Frequently he used a variety of other names for his magazine pieces. Jack seemed to use "Gouger" almost exclusively in the Sydney *Bulletin* until the mid-1920s when he was already preparing to launch his career as an author under his full name. He started to use "I.L.I.", then in 1925 he used "Up North". Also used were "Sniper" in 1928 and "Fingerpoint" in 1928.[16] Jack also used other pseudonyms and had at least two other nicknames. Probably the most curious was the name "Eric". Enigmatically, Idriess wrote in one of his last books that, "*on a selection and locally throughout one district it (his name) was Eric*".[17]

About 1912 Jack earned another nickname (Cyclone Jack) shortly after he began working for the Annan River Tin Mining Company. The "Cyclone Jack" nickname – widely used around the Cooktown district – was (Jack said) first given him by Aboriginal people. He said they called him Watchell (Cyclone) Jacky because he was so tired and sleepy. Jack said the name stuck because he would not work – he said he was "*a great worker when the boss is looking*".[18] Whatever the meaning, Jack said the name was seized on with delight by his white mates because they said he was too slow. His mates said it would take a cyclone to get him moving.

While Jack's names are more than usually mysterious, in 1889 at Tenterfield (New South Wales), the family known as Idriess included a new baby they had recently called Ion. After his marriage in 1888, Jack's father had found difficulty finding work. He could only find a temporary position at Tenterfield. Later during this Tenterfield period Jack's first sister, Ildyce, was born.

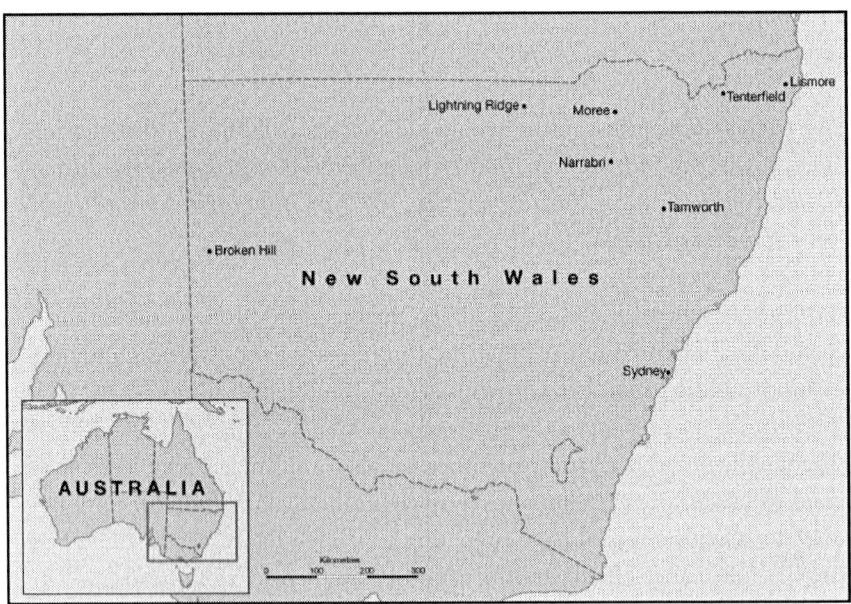

When Jack was about four years of age his father was appointed to a permanent position at Lismore. Jack was too young to remember much about Tenterfield but he had vivid childhood memories of Lismore.[19] Jack lived in Lismore from about four years of age to about the age of seven and remembered the town with affection. He remembered Lismore as a little bush

township with a dusty main street. Jack said the main street of Lismore was crowded with bullock-wagons, cane-cutters, Aborigines, mounted troopers and bearded selectors. There were also the timber-cutters who were felling the great cedar forests. He vividly remembered the big rafts of cedar logs floating downriver to the sea.[20]

The people who most impressed Jack were the raft men; the men who built rafts of logs to float down the Richmond River. His family lived about five kilometres from town on the banks of the river. Their cottage had an attic window from which Jack could see the river traffic. Jack said he would climb up into the attic to watch the raft men with long poles making a raft of pine and cedar logs. He admired their agility jumping from log to log across the river. He could remember huge logs starting down the river with the tide and the raft men jumping across the logs to build rafts with their long poles. He was particularly impressed by the gymnastics of the raft men when they had to break up a log jam. Jack remembered dark tales of cedar pirates along the river and fights between gangs of raft men when hawsers of rival rafts were cut.[21]

When Jack wrote about Lismore his thoughts would turn to the Big Scrub. This was a dense rainforest scrub that grew almost up to the back of the family cottage. He deplored the wanton destruction of the jungle first to get Red Cedar and Pine timber and then to clear the land for sugar cane farms. He bitterly complained of fires burning off waste timber (only the cedar and pine had commercial value) and all the bird and animal life too.[22]

Jack's first experience of Aboriginal people was at Lismore. His life-long interest in Aboriginal people and their customs and skills probably began in Lismore. Jack's deep and abiding respect for Aboriginal people shines through the paternalistic and derogatory language of his time. In thirty-six of his fifty-three books he related his detailed observations of Aboriginal life, customs and spiritual beliefs. He wrote half-a-dozen books specifically about Aboriginal people and in particular, he wrote two biographies of Aboriginal men – two among the few such biographies.

From his favourite attic window Jack said he could see a large Aboriginal camp. Jack said that watching the camp started his interest in the life and customs of Aboriginal people. He told of yarns about the Aboriginal matriarch "Topsy" who ruled the camp when she was sober and was a fighting whirlwind when she was drunk.[23]

One month before his sixth birthday, Jack began school at the Lower Boorie Provisional School.[24] After about eighteen months his family moved to Tamworth. Jack and his family got to Tamworth when he was about eight. However, Tamworth did not seem to stick in Jack's memory as did Lismore. He only wrote a bit about scouting the bush with his mates. After a couple of years the Idriess family moved again – this time to Broken Hill. Jack said he was sad to leave his friend Johnny Allsop who was later to be killed in France.[25] By this time Jack had another sister, Esmé.

Jack said Broken Hill was a major shock for him and his family. He was about ten when they moved there and he remembered the contrasting scenery. From the big rivers and mountains of Lismore and Tamworth Jack wrote about the deadening isolation, lack of water and raging dust storms of Broken Hill. He said it was a dreadful place and living conditions were hard indeed. He remembered flies and dust, typhoid and dysentery and his mother cooking over an open fire.[26] Jack's third sister Katie was born during the Broken Hill period.

Jack always claimed to be a dull scholar at the schools he attended in Broken Hill (the Central Infant and North schools).[27] His first job (over his last school holidays) was at William's the grocer, driving the delivery cart. School holidays over, Jack answered an advertisement for a bottle washer at the Medical Hall Chambers. He found his calling at his next job with the ore buyers White and Hosier's. It was here that he decided he wanted to become a chemist and assayer.

With the blessing of his employers at White and Hosier's, Jack then got a junior's job crushing samples of ore in the Assay Office of the Broken Hill Proprietary Company Limited (BHP). At night he attended assaying courses and in a couple of years he graduated (with honours) as an assayer from the Technical College of the Broken Hill School of Mines.[27]

At the Broken Hill Technical College examinations of December 1906, Jack got First Grade results for the first and second years of Practical Chemistry.[28] Then in December of the next year Jack came up with another

First Grade in the Technical College examinations for practical Chemistry (inorganic).[29]

Jack had his first success as a writer in Broken Hill. It could be argued that his first literary effort to be "published" (in the sense that his essay was on public display) was at the Floral, Industrial and Art Exhibition at the Wesley Lecture Hall, Broken Hill on 25 May 1904. However, whether technically published or not, Jack did win first prize for an Under Sixteen essay competition. In the same exhibition he also won first prize for illuminated lettering and came second in the handwriting competition.[30]

Jack did not like Broken Hill and his negative memories of "The Hill" were capped by the death of his mother which marked the end of Jack's childhood and the abrupt separation from his family.

Jack's mother died in January 1908 when Jack was four months past his eighteenth birthday.[31] Two months earlier Jack had contracted typhoid fever. His mother nursed him at home until he had to be taken to hospital. Jack never saw his mother again because she died of the fever she had caught from him.

Before Jack had even properly recovered from the typhoid fever, his father took him to Sydney and away from his sisters at Broken Hill. Jack never lived with his father again and ever after he blamed himself for his mother's death.[32] It was only a matter of a few weeks after his mother had died that Jack was with his father on a train heading for Sydney.[33] He was sent to live with his step-grandmother Kate Eldershaw the second wife of Jack's maternal grandfather, John Windeyer Edmonds.[34]

As soon as he could, Jack's father returned to Broken Hill. This sudden journey to Sydney so soon after his mother's death might have increased Jack's sense of blame for her death. He might even have thought that his father blamed Jack for the death. Whatever Jack's thoughts at the time (he didn't tell anyone) this journey from Broken Hill to Sydney began his adult life. He was about to start his meandering journey, steadily more up north.

Jack was extremely unhappy (*"unhappy days... longing to get away – anywhere"*[35]) with his time in Sydney. However, this was not because he did not like Kate Eldershaw. He respected Kate for her strong personality and for her success in her business dealings but he was unwilling to submit to her attempts to control him.[36] However, Kate Eldershaw was not the main problem for Jack.

Jack's fragile health was a barrier to his moving on. After all, he had contracted Typhoid Fever only five months earlier and on three occasions during his hospitalisation, doctors were sure he was going to die.[37] He was still in hospital in January when his mother died. Then just six months past his eighteenth birthday, still weak from his own illness and overwhelmed with guilt about the death of his mother, Jack seemed to have been abandoned by his father in Sydney.

Even though he did not much like "The Hill" he was missing his family and his mates. He did not like Sydney which (after Broken Hill) was like an unfriendly, impersonal ant-heap of millions of people. After living his life in rural New South Wales, cities were not for him. So in April, only a month after arriving he had a go at escaping.

For about four or five weeks he sailed as a lamp-trimmer on the little paddle-wheeled coastal steamer SS *Newcastle* and made several trips from Sydney to Morpeth on the Hunter River.[38] On the *Newcastle* Jack remembered dark, cramped conditions populated with countless bugs and fleas. He remembered nights with the ship rolling through surging seas and howling wind (probably damp and cold as well) – conditions that brought on a relapse of the Typhoid. He went back to Kate Eldershaw. Jack said she wore an I-told-you-so expression.[39]

As his health improved, Jack saw an Employment Agency advertisement for station-hands and rouseabouts. He defied the advertisement's blunt requirement that only men with experience should apply and pestered the agent until he was offered a job. For starvation wages that were only a fraction of what he had been earning at Broken Hill and on the SS *Newcastle* he started work on a station near the north-west New South Wales town of Narrabri. Jack did not care about the low wages; he was on his way back to the bush. He avoided Gran's displeasure by leaving a note and slipping out unseen.

So after a couple of false starts, Jack headed off to a life of adventure in the bush. For the next twenty years (now known by his new nickname) he rarely lived in a house. The parallels between Jack's life and George's background are interesting. Although there was a fifteen year difference in their ages, both had set out in their late teenage years to make a new life in the Australian bush.

3

Howick Island

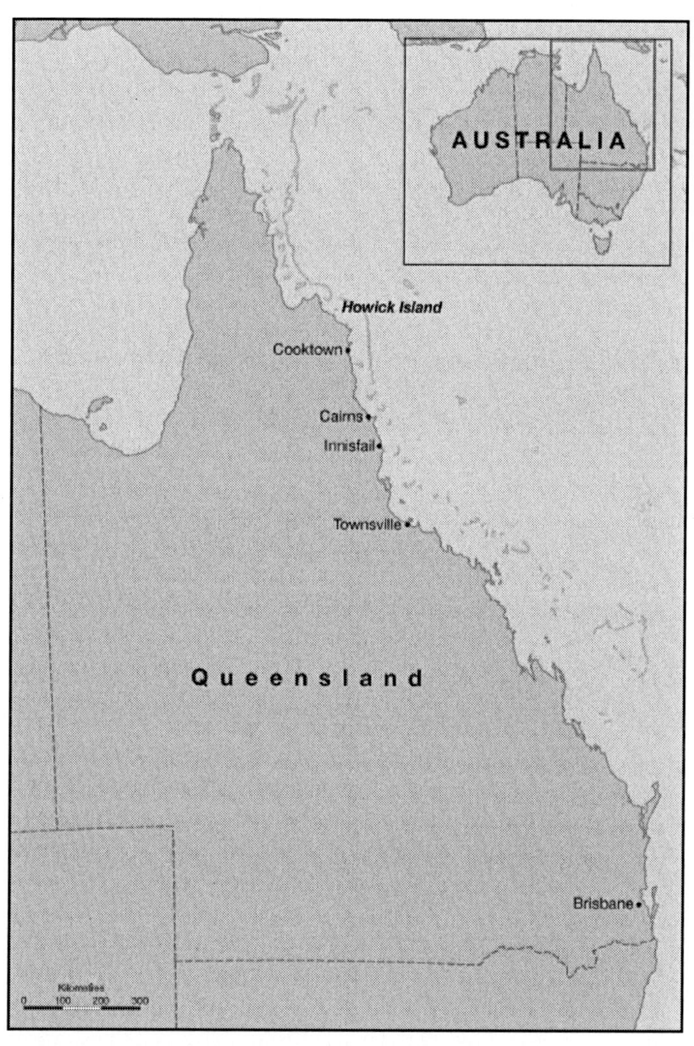

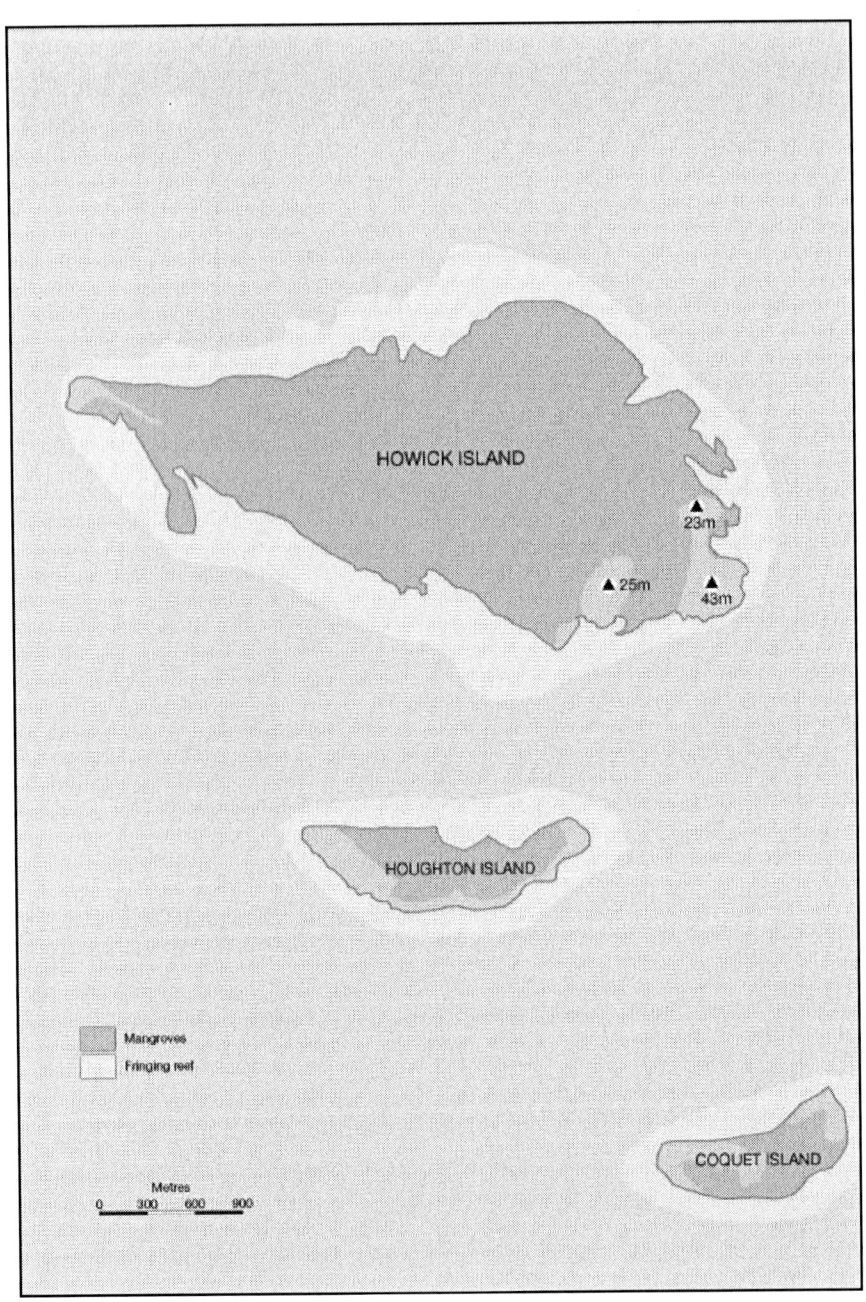

In 1920 Howick Island was known just as Number One of the Howick Group of islands. It is a tiny island with three small hills and about two hectares of permanently dry land. The majority of the island is mangrove forest that is under water at every high tide. George and Jack named the three small hills as the Peak (43 metres), the Hill (23 metres) and the Mound (25 metres). Between the Peak and the Hill there is a small beach. This is where the men made their camp. From the Peak, about a kilometre across a coral reef that is dry at low tide the Mound is just a pile of granite boulders. Howick Island is 100 kilometres from Cooktown in a north-easterly direction and about ten kilometres off the eastern coast of the Cape York Peninsula.

The place of Howick Island in Australian history is tenuous but the sight of Howick (and the coral reefs around it) did influence Captain Cook's decision to discontinue his exploration of the east coast. About 65 kilometres south Cook climbed the Lizard Island peak (now Cook's Look) to search for a route north past Howick to continue his survey of the east coast. Howick Island and the other islands and reefs seemed to be barring his way. After a near-fatal grounding on a coral reef a couple of months earlier, Cook had had enough of coral so he sailed for the open sea. Today's wisdom is that Cook really should have chosen a route past Howick instead of heading out into the open ocean through Cook's Passage.[1]

From the time the island was named it has been mentioned only occasionally in newspaper reports of cyclones, shipwrecks, the occasional death (even a murder[2]) and the island's part in the sad story of Mary Watson[3]. In fact Howick's only real claim to a place in Australian history is that Jack Idriess was stranded there.

There are two good shipping routes past the island.[4] The best lies inshore between Howick and the coast. This is the route that Jack mentions in both *Madman's Island* books. Jack wrote that he saw a large passenger ship (the China boat) using this route.[5] A second, less used and riskier, route is immediately inside the outer Reef (east of Howick).

Howick Island is part of the Great Barrier Reef; a chain of reefs with an inner reef and an outer reef. Further south from Cooktown the reefs are widely scattered and far offshore. Further north the reefs get closer to the coast. From Cooktown onward to the north the Reef becomes the almost continuous barrier that gives it its name. There are only a few relatively narrow passages through the coral to the open ocean.[6] Cook's Passage is less than two kilometres long and under a kilometre wide.

Some islands on the Great Barrier Reef are coral or sand but others (like Howick) were once part of the mainland. The rocks of the three small hills on the island are similar to those on the mainland coast. Most of the island is mangrove forest.[7]

Even getting to the island is difficult. At the right tide it is possible to land in the mangroves at the far end of the island away from the hills but to land where George and Jack made their camp means crossing a reef at high tide. Even at high tide, for much of each year the weather conditions make this crossing dangerous. During the Dry Season the constant South-East Trade Winds blow directly on to that part of the island and during the Wet Season there is a risk of storms or even cyclones.

There are two relatively safe periods just before and just after each Dry Season (say September/October and March/April). Jack found this out for himself. When they arrived on Howick in September the weather was idyllic and they had no trouble crossing the reef. When Jack was rescued in early February the weather was stormy; getting back out across the reef was extremely hazardous. As Jack wrote; their first sight of the island was from outside that dangerous reef. Captain Dan Moynahan in his ketch the *Spray* took them to the island but he could not get close to the shore.

Despite its inhospitable nature the island has had its share of regular visitors. At the right season, mainland Aboriginal people used to island-hop for particular foods. Aboriginal people certainly knew about the spring of fresh water on Howick Island. Curiously, the spring is right in amongst the mangroves and surrounded by salt water. In Jack's opinion it was Aboriginal people who long ago built a collar of stones around the spring to make a small well.

As well as Aborigines it has always been acknowledged that people of several other nationalities had seen the land now known as Australia before Cook "discovered" the east coast and claimed it for England. There have been suggestions that on board the *Endeavour*, Cook had the incomplete charts made by earlier mariner explorers.[8]

The first recorded European to pass by Howick was some twenty or so years after Cook sailed up the coast. An escaped convict, William Bryant,

passed Howick but it is not likely that he would have stopped at this desolate little island. Bryant had a crew of seven fellow convicts. He also had with him his wife, Mary and their two children. In less than ten weeks, they sailed a six-oared fishing boat more than 5000 kilometres from Port Jackson to Timor.[9]

Bryant was a Cornish smuggler transported in 1787 for resisting arrest by customs officers. At Port Jackson, because of his boat-handling expertise, Bryant was put in charge of a fishing boat. In 1790 Bryant spent over five months planning his escape.

Against all the odds, Bryant gathered his crew and waited until there were no ships in Port Jackson to pursue him. Then Bryant, his crew, his wife and children escaped. There is no record of the route Bryant took or the places at which they landed but they must have taken the route inside the Barrier Reef and past Howick Island to get to Cape York. Eventually, crossing the open water of the Gulf of Carpentaria and more open water of the Arafura Sea, they reached the comparative safety of Timor.[10]

The next known European to travel north past Howick inside the Reef was Captain Cripps on his way to India. Cripps sailed his brig the *Cyclops* through the entire unsurveyed inner route but did not record his journey – probably because his was a merchant voyage rather than a voyage of discovery. The irony, of course, is that where the Royal Navy's Lieutenant James Cook feared to sail, a merchant seaman dared and succeeded.[11]

One of the more interesting stories connected to the island is about how it got its name or more specifically, about the man who named the Howick Group of islands: Lieutenant Charles Jeffreys. As he sailed by in 1815 Jeffreys named the Howick Group for Viscount Howick (1764–1845), Charles Grey, the 2nd Earl Grey. It is not known why Jeffreys thought to name a group of obscure, remote and tiny tropical islands after a man who was then a minor politician. Although later he became Prime Minister and by 1815 Charles Grey had been a Member of Parliament for twenty-nine years, his party had held office for only twelve months.

Grey was notable for his unconventional support for abolition of the slave trade and for civil rights for Catholics. He was first elected as a Member of the British Parliament when he was twenty-two. Later he was given the titles

Baron Grey of Howick, Earl Grey and then Viscount Howick. The famous Earl Grey tea is named after him. Charles Grey, Viscount Howick was an interesting man but for the purposes of this story, Jeffreys was the more interesting man but for totally different reasons.

Jeffreys can be categorised with Australia's seafaring explorers along with Cook, King and Flinders. Or he can be classed as a man not only careless of his duties as a Royal Navy officer but also (quite probably) Australia's last buccaneer.[12]

Philip Parker King (a man regarded as the greatest of Australia's early marine surveyors) praised Jeffreys. King said Jeffreys' work did him very great credit.[13] Governor Lachlan Macquarie had a different opinion. After much bitter experience of Jeffreys, Macquarie wrote of him that he was inactive, negligent, vain, conceited and ignorant.[14]

Jeffreys had joined the navy at eleven years of age and served in Royal Navy ships before he received a commission as a Lieutenant in 1805. He arrived at Port Jackson in 1814 in the brig *Kangaroo*; one of two ships Governor Macquarie had hired to transport stores and people between Port Jackson, Van Diemen's Land and Norfolk Island. Jeffreys began earning the ire of Macquarie even before the two had met. Jeffreys took an inordinate length of time (more than seven months) to get to Port Jackson. Macquarie was annoyed and interrogated Jeffreys.[15] From Jeffreys' explanation it is obvious he had begun as he was to continue. He pleased himself without regard to his official responsibilities.

The most charitable explanation of Jeffreys' subsequent behaviour is that he saw himself as a Royal Navy officer and not subject to Macquarie's authority. Macquarie was not so kind. He saw Jeffreys as incompetent, derelict in his duties and incapable of following orders.

Jeffreys botched his first real duty in the Colonies. In May 1814 the *Kangaroo* sailed from Port Jackson. The destination was Hobart Town but sixty-six days later she returned not having reached Hobart Town. Jeffreys blamed storm damage but Macquarie said the damage was minor. Macquarie was so incensed by this effort that he analysed Jeffreys' Log. He found that of the sixty-seven days the *Kangaroo* had been away, Jeffreys had spent only

twenty-six at sea. There is no record of where he was anchored and what was he doing all this time.[16]

When the "minor" storm damage was repaired, Macquarie sent Jeffreys off to Hobart Town with explicit instructions to return as soon as possible. After dropping his cargo of people at Hobart Town, the *Kangaroo* was scheduled to pick up wheat at Port Dalrymple (the original name for the port on the Tamar River at Launceston). In all, Macquarie expected this voyage to take no more than about two months. Jeffreys was away until February 1815: some two or three months late!

Although the *Kangaroo* sailed on time in October 1814 from Hobart Town to pick up the wheat, Jeffreys was not aboard. He had chosen to do some touring overland to Port Dalrymple. It is possible (based on his later and unsuccessful career as a farmer in Van Diemen's Land) that Jeffreys was already searching for land to secure his future after retirement from the Royal Navy. Even when he arrived at Port Dalrymple he refused to load the wheat because he thought it might shift in transit. Macquarie was furious.[17]

Macquarie did not accept Jeffreys' excuses. He questioned Jeffreys' seamanship and his diligence. His first inclination was to send the *Kangaroo* back to England but changing his mind, Macquarie sent Jeffreys and his ship to Ceylon with some of the 73rd Regiment that had finished its tour of duty in New South Wales.[18] Macquarie dispatched Jeffreys from Port Jackson with instructions to sail directly for Ceylon and to get back as soon as possible. Instead Jeffreys began a leisurely tour up the east coast of Australia, naming features as he passed. On what is now known as South Molle Island there is a feature still named Mt. Jeffreys.

When he reached the Endeavour River, Jeffreys was conscious of the fact that Cook's survey of the east coast had ceased once he headed for the open sea through Cook's Passage. So he began a running survey of the Cape York Peninsula from where Cook left off. First he named the Howick Group, named Princess Charlotte Bay and then discovered a new passage through the Torres Strait.

He did not finish his unauthorised expedition until June 1815. It took him well over nine months to sail to Ceylon and back to Port Jackson. When he finally got back Macquarie was more displeased than ever. He fired off another report in which he confirmed his previous opinion that neither Jeffreys nor the *Kangaroo* would ever be fit to render any important service to the colony. Macquarie wrote that he would be returning both ship and master back to England.[19]

In April 1917, Macquarie finally resolved to get rid of Jeffreys. He instructed Jeffreys to go straight back to England without calling at any colonial port. And yet again Jeffreys disobeyed. Using the excuse of needing to repair damage to the *Kangaroo*, Jeffreys entered Hobart Town. In reality he had an illegal cargo of rum on board. Before leaving Port Jackson, Jeffreys had entered into a conspiracy with a bankrupt businessman and former magistrate, Garnham Blaxcell. The deal was that Jeffreys would spirit Blaxcell out of the colonies and back to England. On the way they would call at Hobart Town and deliver the cargo of rum.[20]

But Lieutenant-Governor William Sorell heard about these irregularities. He also had heard allegations about escaped convicts on board the *Kangaroo*. Sorell ordered Jeffreys to proceed to England forthwith. Instead of complying with this direction, Jeffreys just moved down the river into the Derwent Estuary.[21] Sorell sent two boats out on to the Derwent to monitor the *Kangaroo*. Jeffreys objected to this intrusive observation. With an armed boarding party, Jeffreys himself attacked one of Sorell's boats. In an action that could reasonably have been regarded as piracy, Jeffreys assaulted the boat's captain and took him prisoner. He held the captain and crew prisoners on board the *Kangaroo* until the next day.[22]

Jeffreys might have used the captives to negotiate with Sorell. The release of the prisoners might have been some sort of bargaining chip. It is not possible to know but within a week Jeffreys had sailed for England.

Naturally, Macquarie thought he had seen the last of Jeffreys – he expected Jeffreys to finish up behind bars and never to return to the colonies. However, it was not to be so. Jeffreys escaped punishment because of a legal technicality.[23] It is quite possible that Jeffreys' self-promotion of his discovery of a new inner route through the Reef might have influenced the Lords of the Admiralty. Jeffreys remained unpunished.[24]

Nevertheless, Jeffreys did leave the Navy. Later he plagiarised a form of guide-book for people considering settling in Van Diemen's Land.[25] Then about three years later he returned to Van Diemen's Land to set up as a farmer. It is more than likely that Jeffreys had set up his future while he was taking time off from his duties aboard the *Kangaroo*.[26] There is some evidence that he already had his farm under management long before he returned to the Colony.[27] Even so, he failed. Jeffreys died 6 years later.

Jeffreys could have been cartographer or crook or both but one thing cannot be taken away from him. He did accurately chart the inner route past Howick Island and he named a number of features along the coast including the Howick Group of Islands. These names survived despite Macquarie's opposition.

For more than a century after Jeffreys sailed up the coast, the island continued in obscurity. When George and Jack were stranded there it was known still as just Number One of the Howick Group but at some time in the 1920s (before the first *Madman's Island* book was published in 1927) it became known as Howick Island. Since 1989 it has been part of the Howick Group National Park. The island is now managed by the Queensland Parks and Wildlife Service within the Queensland Department of Environment and Resource Management.

4

George and Jack

In 1920 it was about twelve years after Jack had gone bush from Sydney and it was eight years since he arrived in Cooktown. George would probably have been in the area since the early 1890s. After he arrived in Queensland in 1888 as a fourteen-year-old he would have lived around the Townsville area for a while but nothing is known of George's history for the fifteen year period between 1888 and 1903. However, Jack said George had been living for many years in the bush of the Cape York Peninsula[1] so George could have arrived in Cooktown around 1892 when he was about eighteen years of age.

It could have been around June/July 1920 when George, Jack and Dick Welsh met in the West Coast Hotel. In neither of the *Madman's Island* books did Jack give the impression that he and George were anything more than acquaintances. Indeed, Jack implied that their meeting at the West Coast was the first time he had met George.[2] Although there was no explanation in the first book of the way Jack met George, in the second *Madman's Island* Jack related second-hand observations about "Charlie".[3] Jack did not seem to have any first-hand opinions about George but nevertheless, it is likely that they knew each other.

Over the years before they decided to go to Howick together, Jack would have arrived in town occasionally when George was working at the West Coast Hotel. Given that Cooktown was such a small place it is likely that their paths crossed. It is even more likely that George would have known Jack's mate Dick Welsh because Dick was born and raised in the Cooktown area. It could have been Dick Welsh who introduced them – he was certainly at the meeting in the bar of the West Coast when the three of them discussed the possibility of prospecting for tin on Howick Island.[4]

At that time Jack was a wiry sort of bloke about 171.5 centimetres tall and he would not have weighed more than about sixty-five kilograms.[5] Jack's hair (and beard when he grew one) was reddish brown.[6] Eley quoted Jack's sister Ildyce as saying, "*Ion always looked like a countryman. He was about five foot ten inches tall with nice features, the most outstanding of which were his beautiful expressive eyes. I think they were hazel-grey in colour*".[7] Eley corrected Ildyce's memory of the colour of Jack's eyes. She said his eyes were grey and this is confirmed by Jack's enlistment details.[8] Eley wrote that Jack was irritated by the fact that people would mistake the colour of his eyes. According to Eley, when Jack found this mistake he would cross out the offending word and replace it with the word GREY in capital letters.[9]

In 1933 a Brisbane journalist wrote of Jack, "*Idriess, the Australian who made authorship a profitable business, is greying, small, retiring; doesn't look like the conventional adventurer. But he's tough and full of fight*".[10]

George was much the same in height and weight as Jack. In the second *Madman's Island* Jack described George as a tall man but in fact he was just a little bit taller than Jack. He was about 175 centimetres tall and he would have weighed about sixty-four kilograms.[11] Jack said he was sunburned and that certainly would have been true. George had been living under the Queensland sun for over thirty years. At his enlistment George had brown hair and brown eyes[12] but Jack described George as having wild, piercing eyes deep-set under shaggy brows, a face deeply lined and a grizzled black moustache[13] all of which, Jack said, made George look positively savage. However, it must be said that Jack's vivid word portrait would have been aimed at supporting his contention that "Charlie" went mad.

Later, after they had been living rough on the island for about four months Jack wrote that George came across as a grim wild figure, brown body stripped to the waist. He described George dressed in a ragged remnant of trousers but barelegged, barefooted and bareheaded. He said George had a tangled mane of grizzled hair falling to his eyes and that his unkempt grizzled beard mingled with the hairs upon his shaggy chest. Then he modified this derogatory description by commenting on George's "*lithe body in condition of fine strength*".[14]

Jack said George at that time looked like a wild man and later, just as rescue was near, Jack said George's shoulder-length hair was long, matted and hanging down over his forehead and eyes[15] but then, of course, Jack himself would have looked much the same. After months on the island both of them would have scared off any stranger who came face-to-face with them. Even at the end of one month he had a month's growth of beard and hair growing over his forehead.[16] Although his hair was not greying like George's, it was still just as long and matted. Back in Cooktown after his rescue, Jack's mates said he looked like the "wild man of Borneo".[17]

Throughout the second *Madman's Island* there recurs an odd mixture of respect and consistent re-inforcement of Jack's belief that George was mad. By contrast, in the first book George was irretrievably mad and almost constantly at war with Jack.

In the second *Madman's Island*, Jack acknowledged George would have to have been cool and level-headed with grit and initiative to have survived living rough on the Cape York Peninsula.[18] Jack commented on George's physical toughness. With just his fingers, George could lift a live coal from the fire to light his pipe.[19] Jack also said that George could walk with bare feet across needle-sharp poisonous coral that would shred another person's feet.[20]

It is easy to believe Jack's assessment of George as a hard, self-reliant bushman. George would have developed a tough, uncompromising personality firstly from his childhood raised on the wild coast of Devon and then from his experiences arriving in Australia with his teen-age brothers. He definitely would not have been your soft, rosy-cheeked young English boy when he arrived in Australia. He had been raised to be tough and resourceful and he would have needed strong will and a forceful personality to adapt to life in Australia in the late 1880s.

According to Jack, George had spent many years living with Aboriginal people and wandering as a solitary prospector in the far north of the Cape York Peninsula.[21] It is not possible today to fill in the gaps in George's life but because he did spend much of his life in the Cape York wilderness, George must have become one of the few white people to live with and learn from

Aboriginal people. These rare white men had a chance of learning from Aboriginal experts living successfully in a seemingly inhospitable and harsh country.

Jack and George talked about bush food. Jack said some white men can live for a while in the bush and eat bush food in an emergency[22] but Jack said George had actually trained himself to live off the land by eating different things even when he had better quality food. If at first he didn't like some new food, he would keep at it.[23]

These solitary years clearly had an effect on the way that George related to other people. Jack said George was quiet. George agreed. He said that years of life in the jungle brought him to a point where he really preferred to be on his own people. Even so, George also said it was not right to say that he could not stand other people. He obviously saw a difference.

Jack said George lived like a hermit[24] but this was not strictly true. Jack either did not know or overlooked that part of George's history when for a few years around 1903 and 1905 he was cutting cane and working as a locomotive driver (probably on a cane train) around Geraldton (now Innisfail).[25] It is also true that George worked from time-to-time on mining projects with partners[26] and he would have had to be reasonably sociable when he was working as a barman at the West Coast Hotel.

Nevertheless, George prided himself on his ability to be totally self-sufficient and to live without being dependent on the trappings of so-called civilisation. Jack said George lived in a different world. It puzzled him how George could choose a solitary life and (what Jack called) loneliness away from other people.[27]

But then Jack was also a quiet man even though his sort of quietness differed from George's. Jack needed other people around him but he was somewhat detached from them. On several occasions Eley mentions this aspect of Jack's personality: "*Already, though still a youngster,* [twenty] *he appears to have been self-possessed, insular or perhaps seen by the uncharitable as selfish. In fact his personality was gradually developing into one of conscious*

separate existence."[28] Jack's quietness was as though he needed to be close but not too close to other people – as though he was a spectator or like a figure who is always in the background of an old blurred photograph. Another (unnamed) source quoted by Eley said, "... *as a listener sitting back and taking it in while others do the talking, Ion Idriess is in world-championship class.*"[28]

Elsewhere, Eley also quoted Jack's friend Gus Gaunt who revealed, in a 1933 *Sunday Sun and Guardian* interview: "*To even his closest associates Idriess was an enigma, and to get a word out of him was almost as great a problem as to fathom his thoughts.*"[29]

When they talked about this, Jack told George that he longed for his mates and good days travelling in the bush and meeting other people in mining camps and so on.[30] But Jack misread George's situation when he said he did not want the company of other people. George occasionally did *want* the company of other people but he did not *need* their company and he consciously cultivated a protective wall of independence.

For his part, Jack used to talk as though he was afraid of loneliness. He wondered whether a man would become afraid of his own thoughts.[31] George told him he never found being alone was a problem. He had, over the years, simply trained himself to become as independent as possible and to be so self-sufficient that he needed almost nothing from towns and cities.[32]

So both George and Jack were quiet men who were independent and self-reliant. Both were survivors of the First World War. They hardly knew each other before heading off to live for a month on a tiny deserted island. Then they were stranded for months without food and, worse, without tobacco. There just had to be trouble. Was that George's fault? Was it Jack's fault? These are not the right questions.

They were both locked into a situation where the struggle for survival was all that mattered. The real question is why these two taciturn types would have even considered isolating themselves on an island for even a month. With hindsight it is easy to argue the folly of this venture. The other important question is whether it was George's (alleged) mental illness that caused the conflict or something else – perhaps an irreconcilable (and inevitable) personality clash. It is possible that Jack's self-acknowledged laziness and lackadaisical campmate habits might have been the early catalyst for ill-feeling. Jack's mate Dick Welsh put up with Jack's personality. After time on a tiny desolate island George might not have been so tolerant.

Part Two

Stranded

George and Jack on Howick Island
September 1920 to February 1921

5

Planning

Jack said he and Dick Welsh met George in the bar of the West Coast Hotel. As an aside – observant followers of Jack's work will have noted the spelling of Dick Welsh's name. Surprisingly, Jack got the spelling wrong and (no doubt from Jack's notes) so did Beverley Eley.[1] In his books, Jack only occasionally referred to his mate as anything other than just plain Dick. However, when he did use Dick's family name he misspelled it. In the Author's Note to *The Yellow Joss*, Jack refers to Dick as Dick Welch.[2] In *My Mate Dick* he also refers to his mate as Dick Welch.[3] And again – in *The Tin Scratchers* – Dick is referred to as Dick Welch.[4] However, Richard Albert Welsh was born in Cooktown in June 1894.[5]

When they met in the bar of the West Coast, Jack said George had some black stones that had been given to him by a Malay *beche de mer* fisherman Jack called Old Tarquay. The three men agreed the stones were tin and wolfram. Wolfram is more commonly known as tungsten and was used for many years as the incandescent filament for electric light globes. It first became valuable when it was needed to harden steel for armaments prior to the First World War. In the early days when it first became valuable there was plenty for all. In places like Wolfram Camp near Chillagoe lumps of wolfram were just lying around on the surface. Wolfram rapidly rose in price – doubling, doubling and doubling again – a seventeen-fold increase almost overnight. Although George was primarily a gold prospector, Jack and Dick knew about tin and all of them would certainly have known the value of wolfram.

In both of his books Jack said Tarquay had given "Charlie" the stones he had found on Howick when his boat was wrecked and he was marooned on the island.[6] However, there are differences in this account between the two books and in any case, Tarquay's boat was not wrecked on Howick Island.

Tarquay and his *beche de mer* fishing boat the *Sea Foam* did exist. The Log of the *Sea Foam* is still in the possession of Jack's descendants and Beverley Eley saw it.[7] Jack included a photograph of one of its pages in the second *Madman's Island*.[8] Another fact is that in February 1921 (the very month when Jack was rescued) the *Sea Foam* was wrecked but on Hannah Island about 150 kilometres to the northwest of Howick near Cape Melville.[9] Probably drawing on her reading of the *Sea Foam* Log, Eley said Tarquay called Howick "Hammond Island".[10]

Jack said they heard about the tin on Howick from Tarquay but there is a more prosaic explanation of how George and Jack got their information. Tin had been mined on Howick some years earlier and news of this activity had been published in the *Cairns Post* in 1918.[11] This does not mean that Tarquay did not give the black stones to George. Perhaps he found them when he stopped at Howick for water but Jack's accounts of both Tarquay's involvement and of finding the Log of the *Sea Foam* are inconsistent.

There is no mention of the *Sea Foam* Log in the first *Madman's Island* but in that book Jack said Tarquay's boat had been wrecked on Howick two seasons earlier and Tarquay had been stranded for a considerable time on the island before trepang (*beche de mer*) fishermen rescued him.[12] In the second *Madman's Island* Tarquay had been rescued only a week earlier after the *Sea Foam* was wrecked and a crewman was killed.[13]

Apart from the discrepancy in timing, it is inconceivable that a man died and (a) the death was not recorded in the ship's Log and (b) the fatality was not later reported to the police at Cooktown. To illustrate the importance of reporting a death, a seaman called Oosop is buried on Howick and his death was properly reported and recorded.[14] In 1892 there was a murder trial arising from the death of an Aboriginal man on Howick.[15] Even on a remote island, the death of a man is a significant event and it would certainly have been recorded. However, Eley made no mention of the death being recorded in the *Sea Foam* Log and Jack said the Log contained, "*but short words of the sea, or accounts of the sale of* beche de mer, *or orders on Chinese storekeepers for provisions.*"[16]

Even Jack's discovery of the Log does not make sense. In the second *Madman's Island* Jack said he found it some six weeks after he had arrived on the island. Jack said he had found the Log very much the worse for wear, salted and battered but readable.[17] This means that the two men had been living for six weeks on a tiny island when Jack found the diary in plain sight. It is inconceivable that it was not found earlier.

Secondly, it is also almost incomprehensible that this sand and salt-encrusted Log was the only writing material to which Jack had access. There is no valid reason for Jack to use such poor stationery. Although in his early days as a diarist Jack had written on any scrap of paper he could find, by the time of the Howick adventure he had been an amazingly dedicated diarist for at least a decade.[18] He had even kept a formal diary all through the First World War. These diaries still exist. Most, up to 1926, were written in school exercise books.[19] It is just not believable that Jack would have set out on a month-long prospecting adventure to a tropical island without writing material.

It is not known how Jack got hold of the diary but he did not find it "half-buried in sand" on Howick Island in October 1920. The *Sea Foam* was not wrecked until February 1921. Jack probably used the story of Tarquay's shipwreck as a literary embellishment to give *Madman's Island* a dramatic kick-off. The alternative version (getting their knowledge of tin on Howick from the pages of the *Cairns Post*) would not have had the same literary ring to it.

Regardless, there they were in the West Coast Hotel, sometime in the middle of 1920 (perhaps June) and Jack, George and Dick Welsh had heard in some way about the possibility of tin and wolfram on Howick. Even though Dick Welsh was not going with them he seemed to help in the initial planning. There would have been a lot of planning to do. Howick was remote from Cooktown and once on the island there was little chance of correcting any forgotten supplies.

Regardless of how they found out about the mineral on Howick the simple decision in the hotel bar over a beer or two would have been become more complicated. It would not have been at all clear how they were going to get to the island and how long they would need to stay. Jack had seen Howick from a distance and more than likely, George would also have seen the island from the mainland. Jack had been prospecting in that area of the mainland just before the war and George had spent years prospecting the Musgrave ranges inland from Bathurst Bay and Cape Melville. Jack had also climbed the Peak of Lizard Island and seen Howick to the distant north.

They must also have had reliable information about the water available on the island and about a campsite because they went straight to both features when they arrived. The real problem was getting there. For a start the weather was against them. In June the South-east Trade Winds prevalent in the area blow onshore at the end of the island where they wanted to camp. This end of the island is protected by a reef that can only be crossed at high tide and the constant wind at that time of the year makes crossing the reef dangerously unpredictable. Then by December the Wet Season arrives with the threat of monsoons and even cyclones. Their best chance was around September when the trade winds moderate and when the Wet has not yet arrived. There would have been long discussions with the skippers who sailed past Howick while trading along the north coast from Cooktown.

Beverley Eley said it was Captain Dan Moynahan who delivered them to the island.[20] In neither of the *Madman's Island* books did Jack name either the skipper or his boat. Jack just (inconsistently) called the boat a "cutter" and a "ketch".[21] Presumably, Eley got her information from Jack's notes (perhaps from the entries he made in Tarquay's Log of the *Sea Foam*). Jack did travel frequently with Moynahan. For example, in one of his last books Jack wrote that he had travelled up the coast with Captain Dan Moynahan and his mate Doug Hall[22] and Jack heard about the outbreak of the First World War from Moynahan on another voyage up the coast of the Cape York Peninsular.

Jack wrote that all they wanted to do at that time was land on Howick for enough time to check Tarquay's story. This would not have been possible until September. Reading between the lines of the second *Madman's Island* the impression is given that Dick was interested in helping with the planning of the trip but not at all interested in going with them. Dick and Jack would have had some shared gear but Dick would have needed enough to carry on with his own prospecting.

Eventually they agreed to take George's tent and tools. Jack contributed most of the provisions like billy-cans, tinned meat and even some bacon and

the all-important tobacco. They had some cooking utensils and Jack provided some billycans and a kerosene-tin bucket. They took some fencing wire – the bushman's friend – and George took his rifle. They had some clothing but not much because at the end of their stay their clothes were in rags. There must have been other equipment but Jack does not mention it in his books.

"Near-in" Prospecting 1919–1920

According to one of North Queensland's most renowned pioneering bushmen and prospectors, Arthur Sheffield, George and Jack went on a brief prospecting trip around the Cooktown area before they went to Howick. This trip might have been undertaken while they waited for the *Spray*. In 1969 Sheffield (when he was over 90), told Jim McJannett of Cooktown that George and Jack had been prospecting together, "*near in*", prior to the Howick trip.[1] Drawing on Jack's books *My Mate Dick* and *Back o' Cairns* it is possible to consider what might have happened on that prospecting trip. This trip might have given the two men some idea of how their relationship would develop on Howick Island.

Indeed there is a parallel between what Jack told his readers about George Tritton and another solitary bushman – the "Jungle Man" to whom Jack returns time-and-again in his book *Back o' Cairns*. Jack described the Jungle Man (a man with a life style quite similar to George's) with a style that was almost hero-worship.

Nothing is really known about George's life alone in the bush but from Jack's words, he was reputed to have lived alone in the jungle for many years and Jack said he had lived with Aboriginal people.[2] Like George Tritton, the man Jack called the Jungle Man was a silent prospector who wanted no mate – travelling and prospecting on his own.

Jack said the Jungle Man loved loneliness but at times became so desperately lonely that he sought human company. He wrote that the Jungle Man hugged secrets to his heart – "*secrets of the bush that he had learned in loneliness and... years of constant wanderings.*" Jack said the Jungle Man had secrets of humankind and far deeper, hauntingly elusive secrets that puzzled Jack and even made him uneasy.[3] He could use his sense of smell and hearing and his power of observation to an extent far beyond Jack's previous experience.

Asked to compare his senses with those of Aboriginal people, the Jungle Man said Aboriginals had highly developed senses (or perhaps more primitive senses – meaning abilities that so-called civilised men have lost). He said some Aborigines had better senses than others and he included his abilities with those of the others, that is, not as good as some. Like George, the Jungle Man had a reputation for affinity with Aboriginal people. Much of his ability to live efficiently with a minimum of modern equipment was learned from Aboriginal people.

Also like George, the Jungle Man was neat and efficient around the camp and in the equipment he carried.[4] Naturally, he preferred the bush to life in even the smallest of towns.

In his book *Back o' Cairns* Jack first mentioned the Jungle man at page 120 of 310 pages. Jack devoted a chapter to travelling with him then for the rest of the book (right up to the last page) Jack frequently mentioned the Jungle Man. So Jack obviously had plenty of experience living with a man like George. Jack said he filled notebook after notebook with tightly packed scribblings and still he was aware that he missed learning much of what the Jungle Man knew.[5]

It is possible that part of the reason for the breakdown on Howick of the relationship between George and Jack was Jack's unwillingness to help around the camp. However, on that earlier, "near in" prospecting trip George would surely have realised Jack's self-admitted laziness. With this experience it is difficult to understand why George decided to go to Howick with Jack.

In *Back o' Cairns* Jack frequently complained of the Jungle Man's moodiness when he would just stare into the fire or out over the mountains but at one point Jack excused this habit with "*I could be quiet too.*"[6]

Nevertheless, Jack's recurring complaints not just about the Jungle Man's "quietness" but also about his "grumpiness", clearly indicated that the Jungle Man's introspective personality grated on Jack. Often, Jack's under-his-breath reaction was "*… to hell with you…*" and "*… blast him.*" and Jack wrote of himself as feeling grumpy, surly, sulky, glum or grumbling about his mate.[7] But with the Jungle Man Jack took care not to cause any offence for fear of never seeing his "*strange mate*" again.[8]

More pertinently, there were also indications in *Back o' Cairns* that Jack did not pull his weight around the Jungle Man's camp. Although Jack wrote of instances in which he did his share of the camp work[9], often he would still be in bed when the Jungle Man was up and on the move. The Jungle Man would have the billy boiling (and sometimes breakfast cooked) before calling to Jack to get out of bed.[10] This is exactly the behaviour that George encountered even on his second morning alone on Howick with Jack.

However, the laziness that caused the Jungle Man to yell at Jack, "*Billy's boiled. Shake a leg*"[11] was also described in another of Jack's books *My Mate Dick*. Dick Welsh was one of Jack's best mates. Apart from writing a book about him, Jack also dedicated another book – *The Opium Smugglers* – to Dick Welsh. Because Dick and Jack had been mates for years and participated in many prospecting trips, they had a high regard for each other. Nevertheless Dick would frequently get exasperated with Jack's behaviour. Sometimes Dick got exasperated with the poor standard of Jack's cooking.[12] At other times Jack followed his pattern of sleeping in his swag while Dick prepared their breakfast.[13] On a number of occasions Dick became angry at Jack's behaviour.[14] He would get so angry that he would only refer to Jack by his derogatory nickname "Cyclone".[15] Even so, Dick always seemed to have the capacity to forgive Jack's behaviour.

So how did George and Jack get on in that (probably short) "near in" prospecting trip while they were waiting for favourable weather to get them to Howick? Of course there is no answer to that question but surely George must have got some idea of Jack's shortcomings as a campmate. Whatever George thought after the "near in" prospecting trip, he and Jack were on their way to Howick probably in late August. Certainly they were on the island in September so it could have been the last week of August 1920 when the *Spray* left Cooktown.

7

Landing

It might have been just after dawn in the Endeavour River waiting for the tide. Across the river they could have seen Mount Saunders: the tip golden from the rising sun and its base wreathed in the morning mist. On their side of the river, not far from the pubs and businesses of Cooktown they could have looked up at Grassy Hill at the base of which the *Spray* was moored. Captain Cook several times climbed that steep peak trying to spy a route through the treacherous coral reefs through which they were about to sail. But (they hoped) their skipper knew where he was going – Cook did not!

From Cooktown they sailed north along the coast to Cape Bedford. They would have been able to pinpoint and argue about landmarks but as they left Cape Bedford behind they could see little except ocean and the reef in the distance. They were headed for Lizard Island. Captain Cook had explored this far, seeking a way through the maze of coral reefs.

They sailed past Cape Flattery where Cook thought he had succeeded in finding his way through the coral and on to Lizard Island where he climbed the Peak. Cook could see coral reefs to the north and to the north-west the massed islets of the Howick Group. Discouraged, he chose an opening in the Barrier Reef easterly of Lizard Island – Cook's Passage.

Leaving Lizard Island, they sailed north-west. Not far now. Poor Mary Watson sailed this distance in an iron tank.[1] George and Jack had the luxury of safely sitting on the deck of a proper boat. However, even at that early stage (in the second *Madman's Island*) Jack was laying the groundwork for his allegation that George was "mad". On board the *Spray*, Jack described George as,

"... squatting against the mainmast, hatless, peering out over the bows from piercing eyes deep set under shaggy brows. Tall with the beginnings of a stoop, his gaunt frame suggested endurance. His bare hairy chest and brown muscled arms were not those of a weakling. Living like a native had done him no physical harm."

Now those words are not flattering but immediately following that passage Jack paid George the back-handed compliment; "*Mentally he appeared as sane as myself.*" Earlier in the same book he had written of George's "*wild eyes*" and he described George's deeply lined face and grizzly black moustache – all of which (Jack wrote) made George look "positively savage".[2]

So the stage is set for the adventure on Howick Island. These two bushmen, both ex-servicemen, each quiet in his own fashion landed on Number One of the Howick Group of islands in September 1920.

When George and Jack arrived on Howick Island there was a bright blue sky and a slight swell but nevertheless they had to be landed by rowboat across the reef. In his books Jack made light of this landing. It would not have been so easy. For instance, they had a tent, prospecting tools, camping gear and cases of provisions for a month. Jack wrote that all this and four men fitted into just one rowboat for the trip across the reef.[3] This description does not seem likely: more than one trip would have been needed for the *Spray* to land all their stores, camping gear and mining tools.

Just as Tarquay had described, where they landed the beach was only about twenty-three metres long and impossible to see from outside the protective reef. There was a fringe of mangrove trees between the beach and the reef. A few yards above the beach was the one solitary strip of level ground on the island. That was their campsite. Right behind the camp was the main mangrove forest. In the next few days they were to find that there were only a few hectares of permanently dry land on the whole island; all the rest was mangrove and coral submerged at every high tide. In the first *Madman's Island* Jack estimated the area of permanently dry land at two hectares and in the second *Madman's Island* he wrote that there were no more than four hectares of dry land.[4] Either way, it was a very small island.

So there they were left alone to fend for themselves for (they thought) a month. At their tiny campsite there was just enough room for the tent and not much more. They could see the highest point on the island (the "Peak") that rose steeply to the left of their campsite and the smaller "Hill" to their right. In the first *Madman's Island* the Peak is about ninety-one metres high and in the second *Madman's Island* it was eighty-five metres but of course, these are only Jack's estimates.[3] The actual height of the Peak is forty-three metres. In neither of Jack's books did he estimate the size of the Hill but in fact it is twenty-five metres high.[5]

After George and Jack had carried the crates of stores to the campsite they set about making camp before dark. Jack glossed over this bit in his first book. He just said, "*the men rigged their camp, finding plenty of stranded spars for the tent-poles*"[6] but in the second *Madman's Island* Jack made it clear that he left it mostly to George. There were no spars conveniently lying around for tent poles; the tent poles had to be cut. The deal was that George would cut the tent poles while Jack was to get a fire going and put the billy on. First Jack needed to walk to the well for water.

There was a clear track leading away from the campsite between the Peak and the Hill. About a hundred yards further was the mangrove forest. The track led into the mangroves and to the well. So it was not far and it was easy to find. A simple task but Jack was so long getting back that George had cut the tent poles, cleared the site a bit and erected the tent. He also had the fire going. That he had been away so long did not worry Jack at all – he just laughed and commented on George's work.[7]

Without a trace of self-consciousness, in the second *Madman's Island* Jack wrote that he had been just looking around instead of doing his share of the work. According to Jack's own words he had been taking his time sightseeing. He did find the well at the end of the grassy track but instead of getting water and hurrying back, he stood for a while pondering the wonder of a fresh-water well deep into the mud and roots of the mangrove forest. Although neither of the men would have realised it at the time, this was the first of many occasions in which Jack would go off dreaming of birds and trees instead of focusing on the task at hand.

So even during their first hour or so of a projected month-long stay, Jack's tendency to dreaminess or laziness or carelessness around the camp surfaced.

This attitude had been observed by the Jungle Man, Dick Welsh and others – hence the "Cyclone Jack" nickname. Surely, though, George could not have been too surprised after his "near in" prospecting trip with Jack?

Later, sitting around the fire, George and Jack talked about the well and its circular wall of stones. Jack said he thought the wall could have been constructed centuries ago by Aboriginals but more recent visitors had also used it. The more recent visitors could have been the builders of the wall to protect an important source of fresh water. George and Jack talked about this wall and its importance when there is a very high tide. Jack said he thought sea water could rise about two feet deep around the wall. He worried (prophetically as it turned out) about sea water contaminating the only source of water on the island. Later they nearly died when exactly that event really happened.

Just before sunset the camp was set up. George and Jack had collected grass to make comfortable beds and stowed their gear in the tent. Jack highlighted another difference between the two of them. George liked his campsite to be neat and tidy so he had stowed his gear methodically. Jack just threw his stuff into the tent uncaring whether he could readily find things when he needed them.[8] Jack did not say if George was irritated but it is highly likely that he disapproved of Jack's untidiness. These little irritants could have magnified as time passed.

With their camp set up in the late afternoon they sat and smoked a pipe of tobacco then decided to climb the Peak. In the first book Jack wrote that he explored the Peak by himself but in the second *Madman's Island* they both set out to do a bit of sight-seeing from the top of the Peak.[9]

To the west was the last of a vivid red sunset. To the south and west they could just see mountains on the mainland. In the sea all around them they could see dark smudges of other islands. And they got a surprise. They could now see the size of the island and they saw they had another small hill. This thirty-one metre pile of rocks (the Mound) was estimated by Jack at twenty-one metres high (in the first book) and eighteen metres in the second book. The Mound was estimated by Jack to be about a kilometre from the other two hills along a coral reef that was dry at low tide and covered by water at high tide.[10]

Looking to the east from the top of the Peak on that first day, they could see the huge mangrove forest that makes up most of the island. They could see silver ribbons of water that were the channels of sea water running through the mangroves. At high tide all this was inundated.

As the sun set darkness came quickly so they returned to their camp. In Jack's first book he wrote that when he returned from climbing the Peak (alone) George was grumpy and cantankerous. This was not the case in the second book. When both men returned to their camp from the Peak they spent two hours talking together.[11] Jack said they talked about cyclones, mining and life with Aboriginal people.

George told Jack about his experience surviving at the edge of Australia's biggest cyclone.[12] George called it the "Cape Melville" cyclone but more generally it is known as the "Bathurst Bay" cyclone or Cyclone Mahina. The names are irrelevant. It is the destructive force of that huge cyclone that is remarkable. In 1899 the full force of the massive cyclone struck in Bathurst Bay which is at the southern end of Princess Charlotte Bay. Cape Melville is the eastern point of Bathurst Bay.

George was twenty-six years old and prospecting the Rocky River gold field which is about thirty-two kilometres north-east of Coen. It is well inland from the coast of Princess Charlotte Bay and well away from the cyclone centre in Bathurst Bay. The goldfield is in the jungle right up in the mountains. Even so, George told Jack that nine men were killed – pinned under falling trees.

Without warning in the middle of the night the wind flattened the jungle. There was torrential rain and flash flooding in all the creeks. They could do nothing. There was no point in running. In the absolute darkness they could not see anything. They were terrified by the sound of howling wind, rushing water and the crashing of huge trees. George told Jack that when daylight came they found uprooted trees piled more than three metres high.[13] Later George and his mates would have learned they were lucky they had not been closer to the centre of the cyclone.

In Bathurst Bay and around the area (including Howick Island) there were well over 100 pearling boats of various sizes and perhaps seventy of these were sunk. Some ships were refloated but over 400 people were killed. Clement Wragge, then Queensland's Meteorologist named the tropical disturbance he had been monitoring as Mahina – the first time a cyclone had been named. Considering the massive force of this deadly cyclone the name (in Polynesia – a girl's name meaning "the moon") turned out to be completely inappropriate.[14]

A storm surge of thirteen metres was reported as sweeping about five kilometres inland taking with it fish, sharks and dolphins. Rocks were embedded in trees and on Flinders Island dolphins were found fifteen metres up on the cliffs.

Recently the height of the storm surge was disputed[15] but there is a report of an eye-witness, Police Constable J M Kenny, was on patrol in the area searching for a ship's deserter. The patrol had camped for the night on a ridge about twelve metres above sea level and nearly a kilometre back from the beach. A high sand hill was between his camp and the sea. About midnight gale force winds blew away his camp. Later a storm surge covered the ridge to the depth of the constable's waist. This surge continued inland for more than two kilometres. The members of the patrol survived but four of their horses were killed by falling trees.[16]

At the top of the cyclone scale, the Category Five Tropical Cyclone Mahina (with winds in excess of 280kph) was more severe than the category four Tropical Cyclone Tracy that devastated Darwin in 1974. Tracy, with wind gusts up to 250kph, killed seventy-one people.[17] More than 400 people were killed in Cyclone Mahina, including about 100 Aboriginal people who were washed out to sea by the back surge.[18]

It was at the edge of the storm surge of Cyclone Mahina that George felt the wrath of this huge cyclone. George told Jack he had been prospecting for gold in the heart of the Musgrave Ranges. Talking at their campfire on their first night on Howick, George said he had made a living from gold but had better luck with wolfram and he went on to tell Jack about a spectacular failure. On one occasion George thought he had found a mountain of rubies.[19]

He found big rich red stones by the shovelful – thousands of them. He thought he was a millionaire. Quietly, with a few of the best stones in his

swag he hurried through Cooktown and down to Brisbane. However, when George showed them to the first jeweller he told George the stones were garnets. Thinking he was being robbed, George punched the jeweller but jeweller after jeweller gave him the same opinion. George gradually came to accept the fact that he had no rubies, just near-worthless garnets and he was not a millionaire.[20]

In the first book Jack wrote that on this first night the two men talked only briefly and Jack accused George of being in a "cantankerous mood".[21] In the second *Madman's Island* Jack wrote that they talked for hours about cyclones and mining.[22] Perhaps (on their first evening), when they yarning by their camp fire, George and Jack also talked about their adventures when they were just beginning their adult lives.

8

George and Jack as Young Adults

Although there was a fifteen year difference in their ages, both men began adventurous lives around the age of eighteen. George began a life of prospecting and mining in far north Queensland when he was (probably) around the age of eighteen. Jack was just eighteen when, still in Sydney (and sight unseen) he landed six-months employment on a selector's sheep property near Narrabri, about 500 kilometres to the north-west.

After George and his brothers arrived in Townsville in 1888, George's older brother, Washington found a job in Townsville. It is likely (because for all the rest of his working life he worked as a grocer and storekeeper[1]) that he began working behind the counter of a grocery store.

Because George and his other brother William were only fourteen and sixteen, they would have stayed with Washington for some time but at some point they left to work on a cattle property. After George and William left Townsville they lost touch with Washington. In 1893 Washington tried to find them through a missing persons letter in the *Worker* newspaper. In that advertisement Washington referred to the last known address of William and George as "Burleigh Park Station". He asked anyone knowing of their where-abouts to please communicate with W.R. Tritton care of Brodziak and Rodgers, Townsville.[2] At that time Washington was married and he had a little girl, Ivyrine. His wife Olga was pregnant with their second child.[3]

William seemed to have devoted his life to prospecting around the Cairns area[4] so it might be that George had his first taste of prospecting together with William. At some point George must have decided to head further north but there are several gaps in George's history. There is nothing known of him between 1888 (when he was fourteen) and 1903 when he was twenty-nine years of age.

In 1903 George lived in Innisfail (then called Geraldton). He seems to have lived in this area around 1903-1908 when he would have been between twenty-nine and thirty-four years of age. He was working as a cane-train driver (probably prospecting off-season) and in 1908 he declared himself to be a miner at Truxillo, 250 kilometres in a westerly direction from Innisfail.[5] There is another five year gap between 1908 and 1913 but from 1913 onward George lived for the rest of his life in the Cooktown/Coen area as a prospector and miner.

However, the period from 1913 to when George met Jack in 1920 (including his war service) was not enough time for George to attain the reputation of a bushman who had lived like a jungle man with Aboriginal people. He must have had this experience earlier in his life.

At some point, George and William parted company. William stayed around the Cairns area but George headed north. He might still have been thinking about the Palmer River gold although it is more likely that it was John Dickie's find of the Ebagoolah goldfield that drew George to the area.[6] In this hypothetical scenario George might have gone bush around 1892 when he was about eighteen years of age and spent up to ten of those missing years prospecting on the Cape York Peninsula and learning from Aboriginal people how to live in the jungle.

So it is possible that like Jack, George set off on his own at about eighteen years of age. George would have filled in these gaps for Jack just as (in sharing stories of their lives) Jack would have told George about his wandering through New South Wales.

Jack's first job was on a sheep station near Narrabri (New South Wales). Jack said the boss fancied himself as a gentleman farmer and did little actual work except at times like the shearing of his sheep. For the first few months Jack worked from dawn to dusk on any job demanded of him. It was hard work but out in the clear country air his health and strength soon returned after the debilitating Typhoid that nearly killed him and did kill his mother only a few months earlier. Looking back, Jack thought it was the best rehabilitation he could have had.

One thing about Jack – he was never ashamed of his aversion to hard work. He seemed to take a sort of pride in telling people that he would not work. On this, his very first job as an independent adult, Jack's lack of diligence toward his work got him into trouble. Instead of properly milking the boss' cows, he would get some of the milk then let the calves have the rest. There was less milk for the boss but worse was to come. Instead of doing his milking job, he filled the half-empty milk buckets with water. But then there was no cream – so there was no butter for the boss. The boss was no fool. He kept a watch on Jack until he caught him topping up the milk bucket.[7] There was a blazing row and threats of violence.

Then, because he knew Jack had run away from home, the boss said he would report Jack to the police. That was enough for Jack. With a month to go on his contract Jack packed his bag. As night came he crept out along the road that led to the railway siding. There was an anxious moment in the middle of the night when he passed the station homestead. The road lay between the homestead and the men's quarters. The station dogs were snarling and barking. Jack kept low to the ground and from shadow to shadow stealthily worked his way down the road.[8]

Jack said he soon threw away his clumsy bag. He rolled all his possessions into a blanket to make the far more respectable "swag". By the next night he had covered about forty kilometres and (totally without food) he was lucky. Just as he was anticipating a cold and hungry night, Jack came across a campfire. As he drew nearer he saw a tall strongly-built man sitting beside the fire in front of a tent and a partly completed house. The man not only gave him a feed but also got him a job as a bush carpenter's labourer. For the next two months Jack helped build the house. Then as he was being paid-off the boss told Jack he had arranged another job for him.[9]

Jack said he felt he had passed a make-or-break point. He had just had his nineteenth birthday. He had money in his pocket and he was on the road to another job. There was no going back to Sydney. He felt truly independent.

His next job was pure pleasure. His new boss was running sheep on a large selection but his real passion was horses. Against the wishes of his family – they thought he was too old to be breaking-in horses – he and Jack would head off most mornings to yard-up young horses. The boss would quieten the horse while Jack carefully saddled it. Into the saddle Jack was holding on to the reins and mane to avoid being thrown. The boss was actually training Jack at the same time as he trained his horses. For a start he chose horses that he thought would not buck so much. Jack recalled that although he thought he could ride a horse, it was this man who taught him to really ride.[10]

After about four months some travelling shearers told him about Lightning Ridge and how to get there. Opals! Jack heard dazzling stories of fortunes being made in a day. Jack planned to save a bit more money before heading to Lightning Ridge but the boss decided the matter for him. He told Jack he knew that he had run away. The boss' veiled suggestion that if Jack stayed he would be safe from the police did not go down too well with Jack so he left immediately.[11]

On the road again, his luck still held. After only a day or so he was around Collarenabri when he was told about a ring-barking gang. He earned good money with the team of twenty-three axemen steadily killing hundreds of trees every day. Jack said he was tempted by an offer of a permanent place on the team but Lightning Ridge was calling.[12] In any case, by that time he was certain that he could just walk into a job whenever he wanted one.

Following a bushie's directions, he was to avoid the road and follow a track past Six Mile Camp and through the Belah scrub. He was to try to get to Old Tom's camp by sundown. Then (he was told) just go bush directly north and about eight kilometres along he was told he would see a couple of low hills. The bushie told Jack to make for the hills and he would find a track that runs down to the Collarenebri Road.[13] It was not at all easy. Jack had hardly started when he was deep into the creepy Belah scrub.

A dense mass of gloomy trees created black shadows eerily pierced by filtering moonlight. There were no familiar sounds, just the shivery whispery sighing of wind in the trees. Jack was feeling his way from tree trunk to trunk, deciding whether to camp when a piercing howl had him crouching in fear behind a tree. The howl was closely followed by maniacal laughter and

another wild shriek. Breathing painfully Jack was trying to settle his nerves. Out of the silence came a shouted command and a burst of song. He realised he must be close to Old Tom's camp.[14]

Jack had been warned about the mad boundary rider Old Tom and his fiercely protective dog. Jack saw a chink of light in the blackness and crept closer to hear Old Tom in character as "Henry V" leading his troops into battle. One surreptitious glance at the fierce cattle dog (Old Tom's audience) and Jack decided against making himself known. By starlight he headed off along the bushie's directions.[15]

At daylight next morning there were the two hills, hazy in the early morning light as the sun rose over the tree tops. All day he walked through the trackless bush and congratulated himself on coming out just where he wanted to be on the Collarenebri Road. Just on dusk he spotted a campfire and even as he approached he could hear a man by the fire talking to himself. Jack soon found that if no-one was around this man would continually talk to birds and animals and of course, himself. He talked incessantly.[16]

Glad of company, Jack camped with him and the man talked far into the night. The problem was that with first light there he was again – still talking. That day as they walked on toward Collarenebri the little man talked enthusiastically about wild schemes for making their fortune. All he needed was a good mate.

In amongst all this chattering, Jack was amazed by an incredible talent – the little man could really communicate with birds and animals. When they stopped for a rest under a big old tree he called up crows by the dozen. The man said that, camped alone, he could sometimes have dozens of birds around him. He showed Jack some clever gadgets he had made that let him imitate the yelp of a fox or the howl of a dingo. He said he could also call up snakes and even frogs. The deal he put to Jack was that he would call up dingoes and Jack would shoot them. There'd be a fortune in it. They'd have a first class set-up with their own horses and cart and everything. That sounded good to Jack but the little man kept chattering on until Jack could take it no longer.[17]

Although Jack was aiming for Lightning Ridge he deliberately took a

detour because a Collarenebri storekeeper told him there was more work in that direction. Along the road from Collarenebri to Walgett he came across the Woorawadian Station. Around August of 1909 he arrived at the Station with a couple of sundowners who were looking for food but Jack was looking for work.

By that time he had been continuously employed in the bush for a year so he was able to tell the boss he was used to bush work – fencing, breaking-in, ringbarking, mustering, burr-cutting, crutching sheep and so on.[18] Jack said he was told to work with a man called Scotty. Their job was to drive a poison cart to poison rabbits.

The poison cart was just a drum on wheels. As a horse pulled the cart the wheels turned a mechanism in the drum and a tube with a pointed tyne dug a furrow in the earth. Pollard and bran laced with phosphorous dropped into the furrow and a trailing chain covered it up. Rabbits died a dreadful death as the phosphorous burned their insides. Jack was sorry about this but he was really appalled about the birds he also killed. His solution was to mix the bran, pollard and molasses but without the poison. So he was feeding the rabbits not killing them. From then on, so they would not lose their jobs, Scotty mixed the poison.[19]

Jack worked on Woorawadian for about five months – up to his twentieth birthday. He worked on jobs other than the poison cart. Woorawadian was a sheep station and Jack described marking the lambs, snipping a piece from each lamb's ear, castrating the males and docking their tails. Hot, dusty, noisy, high-pressure work and Jack marvelled at the effort needed to get meat to market and on to dinner plates. The last job Jack was given before he headed off again toward Lightning Ridge was riding the station boundary fences, checking and repairing them if necessary. Totally alone, Jack thought his job was great in the daytime but he did not like the lonely nights.[20]

Over their campfire, talking late into their first night, George and Jack might have shared these early life experiences but eventually they went to bed thinking about the next day's prospecting.

9

Prospecting September 1920

Next day was going to be the test of Tarquay's claim that he had found the black stones just lying around on the beach. Even today the black stones can be found on the island but more significantly, it is still possible to see the site of the bush kiln made by George and Jack to help them crush the ore. It is close to the campsite and the intensity of the fire has scorched the ground to the extent that vegetation no longer grows where the kiln had been located.[1] On Day Two George and Jack were optimistic.

George was out of bed before daylight and got the fire going. Jack was still asleep. In Jack's first book he does not mention that George already had the billy boiled and breakfast cooked before George called Jack to get up. George was not happy.[2]

It did not at all worry Jack that he got out of bed reluctantly (and only then because George called him). He just grinned and admitted that he was lazy. At that time, it is true, Jack's laziness did not matter too much.

After breakfast they started looking for tin. The first results were not at all promising. By midday they had prospected the whole of the tiny beach. They were disappointed with the poor showing. During the afternoon they dug a line of holes but still found almost nothing. The little beach was a failure.[3]

That night (in both *Madman's Island* books) Jack wrote that they talked about Tarquay's black stones and the almost inevitable disappointments of prospecting. Jack seemed to take this disappointment to heart. He had thought there would be a quick and easy result that would make them both rich. George was more hopeful, probably because he had more experience of prospecting and its vagaries.

However, in the second *Madman's Island* Jack also wrote about he and George sitting around the campfire smoking their pipes when they resumed a topic of conversation they had begun the previous evening – about Aboriginal people and their customs.[4]

10

Aboriginal People

Howick Island has never been conducive to permanent settlement but at the right time of the year Aboriginal hunting parties would start north of Howick and island-hop down at least as far as Lizard Island. They would visit known sources of valued food like seabirds and turtles and their eggs. At Lizard Island the lizards were their target.[1] Jack acknowledged these visits in describing the well on Howick Island.

In this book, with no disrespect intended, the terms "Guugu Yimithirr" or "Guugu Yimithirr people" are used and the Guugu Yimithirr tribe are referred to as a discrete people. However, the name has many different spellings and the Guugu Yimithirr people never comprised a single entity. Even today, a Guugu Yimithirr Story-teller, Wilfred (Willie) Gordon, says he can only speak of his own Nugul clan and cannot under any circumstances speak of the other clans surrounding his own country.[2] There are thirty-two clans of the Guugu Yimithirr tribe. In one sense the tribe is based on its language; the title Guugu Yimithirr refers to language grouping. Guugu means "speech" (or perhaps words or voice) while "yimi" means "this" and "yimithirr" means "this way". There are two language divisions: Thalun-thirr (seaside) and Warrgurgaar (outside).[3]

Traditionally, the Guugu Yimithirr occupied the area north of the Annan River and their territory extended along the coast to Princess Charlotte Bay and the offshore islands including Howick Island.

There is, of course, no written record of the lives of the Guugu Yimithirr prior to Cook's contact with them but a clash of cultures was inevitable. At the core of the clash was land ownership. Stanner (1968) said that,

> *"No English words are good enough to give a sense of the links between an Aboriginal group and its homeland. Our word 'home', warm and suggestive though it be does not match the Aboriginal word that may mean 'camp', 'hearth', 'country', 'everlasting home', 'totem place', 'life source', 'spirit centre' and much else all in one. Our word 'land' is too spare and meagre ... to put our words home and land together into homeland is a little better but not much ... the Aboriginal would speak of 'earth' and use the word in a richly symbolic way to mean his shoulder or his side."*[4]

For the Guugu Yimithirr people spirituality was an integral part of their existence. It encompassed all living things and also the land around them. They saw their very survival as being linked with spirituality. Today Willie Gordon says spirituality and survival are about sharing knowledge and values: *"If we share our cultural values with each other, then we will all be much stronger for life's journey."*[5] Central to the beliefs of the Guugu Yimithirr is the idea of Guurrbi – a time or place made sacred by the Rainbow Serpent. Willie Gordon says his personal Guurrbi – his place for reflection – is where he takes tourists to see (only part of) the sacred sites and traditional rock paintings. Some parts of this country are off-limits.

It is not clear how the tribes of the Guugu Yimithirr were governed nor how they related with one another but in a land rights claim over the Gove area of the Northern Territory, Blackburn J (1971) found:

> *"The evidence shows a subtle and elaborate system highly adapted to the country in which the people lead their lives, which provided a stable order of society and was remarkably free from the vagaries of personal whim or influence. If ever a system could be called 'a government of laws, and not of men', it is that shown in the evidence before me."*[6]

Following a shipwreck of a Lutheran missionary in 1885 a Lutheran Mission was established at Cape Bedford. The Cape is clearly visible from the Peak on Howick Island. The mission was established for the local people to mitigate the adverse effects of continuing prospecting after the Palmer River gold rush. In 1949 the community moved to better land and water at what is now Hopevale about twenty kilometres inland and forty-six kilometres north

of Cooktown. It was German Lutheran missionaries who first wrote down the language but the name came to acquire many spelling variants, some of which are: Gogo-Yimidjir, Guugu Yimidhirr, Kukuyimidir, Koko Yimidir, Kuku Yimithirr, and Kuku Yimidhirr.

The first recorded contact between the Guugu Yimithirr and outsiders was when Cook arrived in the Endeavour River with his crippled ship and a large party of Europeans.[7] From contemporary journals it is evident there was goodwill on the part of both the Aborigines and the newcomers. The Guugu Yimithirr people were definitely curious and also friendly and tolerant. For example, they allowed these strangers to explore their country and collect food. It was not until there was a dispute over some turtles that there was conflict and even then the conflict was minor (relative to the later conflict between the Guugu Yimithirr and the Palmer River prospectors). Even when one of their men was slightly wounded by bird-shot, the Guugu Yimithirr people still forgave Cook and his party. Good relations were restored before Cook left.

Goodwill was also evident in the attitude and behaviour of Cook and his party. Cook was renowned for his attempts to understand and relate to native people. He deliberately tried to make contact and communicate with the Aboriginal people he saw when he first landed at what is now called Botany Bay. The journals of Cook Banks and others record their unsuccessful attempts to establish contact. Despite being unimpressed with the standard of Aboriginal canoes and dwelling Cook made the (often quoted) observation:

> *"They live in a tranquillity which is not disturbed by the inequality of condition. The earth and sea of their own accord furnishes them with all things necessary for life. They live in a warm and fine climate, and enjoy every wholesome air, so that they have very little need of clothing. In short, they seem to set no value upon anything we gave them; nor would they ever part with anything of their own. This, in my opinion argues that they think themselves provided with all the necessaries of life."*[8]

Banks seemed to have a similar attitude. Later, at the Endeavour River, Banks wrote of sighting the first Aboriginal fire:

> "*At night we observed a fire ashore near where we to lay which made us hope that the necessary length of our stay would give us an opportunity of being acquainted with the Indians who made it*".[9]

Later, on the 18 June 1770, Banks referred to the Guugu Yimithirr who visited the *Endeavour* as "*our very good friends*" and Cook, throughout the stay, continued with his demonstrated respect and empathy for indigenous people. It was not so with the next known visitor; Phillip Parker King.

In June-July 1819 King stayed a fortnight on the Endeavour River and twelve months later he anchored again in the Endeavour River. King's attitude from the start was one of suspicion and readiness for conflict. His very first task was to burn all the grass around his camp to avoid, "*a repetition of the revengeful and mischievous trick which the natives formerly played Captain Cook*". On each of these visits there was conflict with the Aboriginals.[10]

Some thirty years later, land exploration in the area began with Edmund Kennedy who in 1848 was the first European to penetrate the country around the Palmer River. The Guugu Yimithirr defended their country fiercely against this intruder. At one point the Guugu Yimithirr fighters tried to burn out the Kennedy party just as they had done to Cook. Kennedy was harassed by the Guugu Yimithirr right to the end of his journey. Kennedy was killed by a spear almost within sight of the ship waiting to take him home.[11]

It was the influx of prospectors and miners that really started the deadly conflict between the Guugu Yimithirr and the new arrivals. In this unequal clash of culture and technology, the Guugu Yimithirr lost their land and much more. The first large-scale conflict was the one-sided fight at Battle Camp.

Five days after the steamer *Leichardt* landed in the Endeavour River on 24 October 1873, the first party of about 130 men set off for the Palmer River. On 4 November the party camped on the Normanby River. The party was either warned by the existence of fresh tracks or they might have been conscious of an attack on the party of William Hann the year before.

Whatever the cause of the party's wariness, the outcome was a barricade of carts, equipment and saplings around the camp.

Before dawn next day a barking dog alerted the camp. About forty Guugu Yimithirr men charged the barricade but there could have been about 500 Aboriginal people at the site of the battle. About half of the men charging the barricade were killed immediately at the very first volley from the defenders at least some of whom had the modern breech-loading Snider rifles. A second volley saw the remainder of the attackers at the barricade to engage the party in hand-to-hand combat. A third volley caused the Guugu Yimithirr men to retreat but they still threw spears from cover. When the Guugu Yimithirr fled the battlefield they were pursued. Some of the Aboriginal fighters were hunted down and killed by troopers but others escaped and panicked the party's horses. The hobbled horses were recovered unharmed and the party continued to the Palmer. This conflict gave the location its name – Battle Camp.[12]

The deadly fight at Battle Camp was not the first clash between the Guugu Yimithirr people and non-Aboriginal intruders but that brief incident changed the way the Guugu Yimithirr fought. At Battle Camp they found a massed attack was useless against the intruders' guns. From that point they used hit-and-run guerrilla attacks on people travelling to and from the Palmer goldfield. Over the years these attacks were documented, dramatized and often exaggerated in the newspapers of the day and subsequent literature. There were associated (and sometimes disputed) claims of cannibalism.[13] Jack repeated some of these stories.

However, even though Jack had arrived in Far North Queensland in 1912 and he had solid experience of Aboriginal people through his part-Aboriginal mates Norman and Charlie Baird and his other good mate Dick Welsh, his thoughts on Aboriginal people barely get a mention in the first *Madman's Island* book. In his first book, Jack attributed the building of the protective stone collar around the freshwater well on Howick to "*Aborigines in long past centuries*". At no point in that book did he give any indication of life-style and customs of the Aboriginal people he had met around Cairns and Cooktown.[14]

In the second *Madman's Island* Jack wrote that he and George talked about Aboriginal people. Now, whether Jack wrote from his diary or from memory or whether he padded the second book with his own reminiscences, George was reputed to have lived with and as a member of Aboriginal tribes.[15]

On Day One they talked about surviving on bush food and some of George's experiences of living with Aboriginal people. Although the overall tone of Jack's writing was not derogatory, he used terms that few people would use today. He wrote of, "*the animal scream of a buck awakening with a spear between his ribs*" and he used the words "nigger" and "lubra" with no trace of self-consciousness.[16]

Almost at the end of their time on Howick Island, Jack wrote about George's experience of Aboriginal spirituality. These views could have been accurately attributable to George or they could be Jack's own ideas. There is no way of knowing but certainly Jack's experiences of and attitudes toward Aboriginal people are well known through his books and acknowledgement must be made of Jack's writing about Aboriginal people.

Jack wrote a tremendous number of words about Aboriginal people. Out of Jack's fifty-three books, thirty-six contained substantial information about Aborigines based on his own observations. He also repeated stories he had collected in his travels – stories about the life-style and customs of Aboriginal people.

Jack also wholly devoted many books to stories and observations about Aboriginal people (including – well before most other writers – biographical material on notable Aboriginal men). To analyse the thirty-six books in which Jack writes extensively about Aboriginal people would be a monumental task but there are four points that must be emphasised.

Firstly, from today's viewpoint some of Jack's language is offensive. Secondly, his writing displays attitudes that are not today acceptable. These points are indisputable. Thirdly, however, taking a broad view of Jack's work he appeared to have genuine empathy for Aboriginal people that went well beyond mere curiosity. He seemed to have a genuine interest in Aboriginal people, their lives, their customs and their beliefs. Lastly, it is self-evident that

Jack's language and attitudes were products of his times – he did not create the (now offensive) words or formulate the (now objectionable) attitudes.

From today's perspective it is easy to find (in Jack's writing) evidence supporting an allegation of flat-out negative racism. For example, in 1943 Jack wrote, "*Never become too familiar with the Abo, but treat him in a friendly way, and leave him with the impression that you are a friend of him, and he of you. Then, should an opportunity occur later, he will do anything for you.*"[17] Many other examples can be found in the Sydney ***Bulletin's*** "Aboriginalities" pages. Reading Jack's regular contributions would make today's reader most uncomfortable but he did not edit or publish "Aboriginalities" and he was not the only contributor.[18] The fact is that the language Jack used and the attitudes he held were those of his times.

For instance, Jack's language was no different from the 1926 recommendation to scientists visiting Aboriginal communities. Miller (2002) said that among other advices, scientists were advised to act in a firm (but considerate) superior and masterly way. The visiting scientist was warned against walking immediately in front of an Aborigine (or even turn his back on one) in case an "*irresistible desire*" might overcome "*the unstable mind of the savage*" to use a tomahawk to attack the scientist. The visiting scientist was advised to extend friendship while showing no signs of timidity but nevertheless to always carry a revolver on his belt.[19]

This is not to say Jack's work should be accepted uncritically but his work should not be ignored because of the language of his day. At the least, Jack's work should be examined in the context of Aboriginal studies and Australian literature. Jack's books should be important today for exactly the reason that derogatory language and dismissive attitudes toward indigenous people were part of his contemporary society.

Modern readers should take into account that Jack was writing about a period before and after the First World War when Aboriginal tribal culture was dying under the influence of contact with white culture. Jack grieved for the loss of this culture and deliberately set out to develop a deeper understanding of it. Despite the condescending language, Jack's books

demonstrate his genuine quest for understanding of the lives and culture of Aboriginal people. He was an honest observer of a lifestyle he could see vanishing during his life-time.

Jack's books deserve consideration when Australian literary perceptions of Aboriginal people are under discussion but there are academic commentators who simply ignore his work. It is astonishing that writers wrote such titles as, "*The Treatment of the Australian Aborigine in Australian Fiction*" or "*Literature and the Aborigine in Australia, 1770-1975*" while completely overlooking dozens of Jack's books about Aboriginals some of which sold thousands of copies.[20]

Of course there was no way George and Jack could have talked about all these ideas – Jack was still years away from writing even one book. Nevertheless the two men did talk about Aboriginal people during their first couple of days on the island. Then they went to bed thinking about the next day and the possibility of exploring their domain.

11

Exploring
September 1920

The next morning, their third day on the island, began well. George got up before daybreak to get the fire started. Their world was at peace. There was none of the stormy crashing surf they were later to endure. He would have only heard the murmur of the sea in the pre-dawn darkness. George would have lit his pipe and watched as tiny flames grew into a decent fire that would give him the coals to cook breakfast. Sparks from the fire would have been swirling up into the dark sky. There was the promise of a clear sunny day. At that time of the year, before the weather turns hot and humid for the Wet season, there can be days that are sunny yet cooled by a light breeze.

As the sun was just about to rise when the fire had burned to a mass of red coals. George prepared a damper mix of flour and water with just the right amount of baking soda, salt and some raisins. This mix was in George's old iron camp oven and into the coals just as Jack woke up.

They seemed to have reached an understanding about cooking (perhaps from the experience of the "near in" prospecting trip) and it was George who did the cooking. Jack wrote that he hated cooking and so was not good at it.[1] Not only in *Madman's Island* but in other books Jack wrote of his reluctance to cook and (consequently) his poor cooking skills.[2]

Jack washed the breakfast dishes. Well, rough enough – he wrote that the technique was to put the tin plates and mugs into a prospecting dish, pour hot water over them and then lay them out on the grass to dry.[3] Simple enough but it's not a method to be found in good housekeeping manuals.

After breakfast, starting out across the reef with the clear blue of the sea on their left-hand-side and the dark green mangrove forest on their right they strolled casually along. George and Jack were talking amiably. They stopped to examine the queer things living in pools of water left by the outgoing tide. Jack was amazed that George was able to walk across sharp coral in bare feet but as George said; from years of walking barefoot in the bush the soles of his feet were as tough as any leather.[4]

In the first book, even at this early stage of their stay on Howick, Jack wrote little about any discussion between the two men. In the second *Madman's Island* Jack wrote that he and George talked about the animal and plant life they saw[5] but prospecting was uppermost in their minds. Even on this sightseeing tour they were carrying prospecting tools.[6]

So it is quite likely (strolling across the reef yarning in the sunshine) that they would have shared their experiences of prospecting. George would have told Jack about gold prospecting and there is no doubt that Jack would have told George about opal mining at Lightning Ridge.[7]

12

Lightning Ridge 1909–1912

About the time (around 1908 and 1909) when Jack was working his way through country New South Wales toward Lightning Ridge George was working as a miner at Truxillo Mine in Central Queensland but for some years he was working as a locomotive driver in Innisfail (then Geraldton). He was probably driving a sugar-cane train and prospecting in the off-season.[1]

Unfortunately little is known of George's life around this time but from his book *Lightning Ridge* it is possible to track Jack's movements (even so, it is difficult to accurately ascribe dates to his adventures).

On his way to his first visit to Lightning Ridge in 1909 Jack said he had plenty of money in his pocket when he left Woorawadian Station where he had been working as a rabbit poisoner and boundary rider.[2] He had just celebrated his twentieth birthday when he set out to trudge the eighty kilometres to Lightning Ridge. He took out his first Miner's Right in September 1909.

Jack's first sight of the Ridge was of a wide flat area mostly cleared of timber with a maze of dumps topped by windlasses. He felt he had finally found a chance for real freedom. Now he had no boss: at last he could work and think for himself.[3] He could see a few bark huts but mostly there were tents everywhere. He remembered hundreds of men toiling at digging shafts in hope of finding the elusive opal. These "gougers" were kept going by the sight of the results of a lucky find. Jack described his first sight of opal as pinpoints of orange and green flashing in the light of the fortunate miner's fire. He also remembered seeing the hugely valuable black opal on display at the Lightning Ridge store.

On this first visit to the Ridge, Jack had no luck at all – at least not with the mining. The real value of this period was when he met Tom Peel, a man reputed to be a disgraced lawyer.[4] It was Tom Peel who suggested to Jack that he begin a professional writing career. Jack had told Tom he wrote essays at school. So he probably told Tom that he had long ago won an essay competition in Broken Hill.[5] According to Jack, he was a reluctant starter; he said Tom had to bully him into writing an article for the Sydney Mail.

Under the name "Miner", Jack's first published article appeared in the Sydney Mail.[6] It describes life on the diggings under three headings: "*Opal Mining at Lightning Ridge*", "*Snipping and Classifying*" and "*Grinding the Potch*". To Jack's surprise he received three guineas for the story.

Jack's riches from his first stay at the Ridge came from his efforts as a writer rather than his luck as a miner. He said he and Old Tom Peel did not find a solitary stone even though they were digging in the very heart of the only black opal field in the world.[7] Even with this lack of success, the people Jack met at Lightning Ridge and the excitement (others had) of finding opal created a lifelong interest in prospecting and mining. When Jack was nearly broke and leaving the Ridge, he swore to Tom Peel that he would be back to strike opal. Jack did go back to the Ridge but first he needed a job.

Even before he left Lightning Ridge Jack got a job as a horse-tailer in a small droving outfit.[8] A horse-tailer is responsible for keeping the horses in good condition so that the sheep or cattle could continue travelling about eight kilometres a day along the stock routes. These stock routes were frequently eaten bare of grass but in the privately owned paddocks on either side there was usually plenty of feed and water. As the drovers passed by they were watched carefully by suspicious cockies – the farmers or graziers who needed the grass for their own stock. It was Jack's job to make sure the horse got some of the sweet grass on the cocky's side of those fences.

Late at night he would creep out of the drover's camp and lead his mob

of horses back along the track to a spot he had noted during the day – a spot where he could stretch the top fence wires down so his horses could step over the fence into the paddock. As daylight approached he would be ready for a quick departure back over the fence, leaving as few traces as possible. But, inevitably, one night he got caught.

Many years later (when Jack was by then a respectable and popular author) he was asked to speak at the Narrabri Town Hall. After Jack's speech an old retired cocky shook Jack's hand and told him he remembered when Jack stole his grass. The old cocky told him that if he could have caught him stealing his grass he would have tanned Jack's hide with a stockwhip.[9]

That particular cocky did not catch Jack but eventually another angry stockman did catch up with him – and he did have a whip. Twenty years later Jack could still remember the bite of the stockwhip on his fleeing backside. He said no-one could understand the pain of a hiding with a greenhide whip unless he has actually experienced it.[10] Jack had stretched the fence wires to let his horses into a good paddock of feed but when (just on dawn) he went back to get them they had strayed too far into the paddock.

By the time Jack found his horses it was daylight and the stockman had discovered the intruder. A leisurely stroll back to the stretched fence panel soon became a full-on gallop when Jack glanced back to see a man with a whip racing out of some trees on a big grey horse. With the whip flying and cracking, Jack zig-zagged on his pony to give the other horses time to cross the fence. By the time they were all out again on to the stock route Jack was forty-five metres behind them. It was time to go.

With his tail feeling like raw steak, Jack raced for the fence. The only means he had of limiting the whip damage being done to him and his pony was a branch he had snatched from a tree. Holding it behind him he was able to snag the whip sometimes. The stockman on the big grey horse was too fast and the stockhorse stayed about six metres behind and following Jack's pony's every twist and turn. Time and again the whip bit at Jack and his pony. Then he saw another problem. As the last of his horses hurried over the fence, the stretched wire sprang back into place leaving Jack on the wrong side of a five-wired fence and a demon with a whip right on his tail. Jack threw the branch back in the face of the grey stockhorse and it swerved. It gave him time to take a running jump at the fence. At a full gallop the pony just cleared it.

Back in camp (and before the angry stockman arrived) Jack explained all this to a sympathetic but unsurprised boss-drover. Nursing his cut backside Jack rode off (standing in the stirrups) to hide in some trees. He left the boss to deal with the stockman by denying any knowledge of Jack's grass-stealing. Despite this painful experience the droving job was good for Jack's finances. When the job cut out he had enough money to last him for twelve months of prospecting back at the Ridge. Every week twenty-five shillings had been carefully saved in a tin box made for tobacco.[11]

So Jack rolled his swag and started walking back toward Lightning Ridge – a casual swagman mixing with the professional Sundowners who lived permanently on the road. Jack's luck held. Along the way he got a few weeks' work as a rouseabout in a shearing shed.

Jack's second stint at the Ridge probably from around March 1911 was productive. Beverley Ely said, "*.... he was hardly ever off opal.*"[12] Eley said Jack was at the Ridge on this second visit for two years but it was probably more like twelve months. Jack was at the Ridge in March 1912 because he described the floods of 1912 in an article for the Sydney *Bulletin*. His next *Bulletin* articles are from Queensland and relate to prospecting for tin and gold.[13] It was probably around March 1911 (when Jack was about six months short of his twenty-second birthday), when he arrived back at Lightning Ridge.

His luck immediately improved. Because his mate Tom Peel was working with another gouger, Jack had a go at an abandoned mine and struck opal. In a week he had chipped out opal valued at nearly £100. He said he would need to work for about two years at a labouring job to earn so much money.[14]

It was on this second visit to the Ridge that (again under Tom Peel's pestering) Jack began to properly analyse the *Bulletin's* "Aboriginalities" column. He submitted dozens of his paragraphs before Tom one day came hurrying with the mail. There in the "red rag" was one of his "pars" under the pen-name "Gouger". So Jack became "Gouger of the Bulletin".[15]

Jack and Tom had partnered up again and soon experienced all the highs

and the bitter lows of a miner's luck but generally they were on opal for most of 1911 and into 1912. Their good luck turned when they took on an older man (Old Dad) as a partner. Not only did Dad talk about their most recent opal find (so that the opal was stolen in the night) but Dad also pegged the claim wrongly. Worst of all, Dad managed to die mysteriously and Jack found himself suspected of murder.

The suspicion arose because Jack had sold a nice little parcel of opals for £72 to be split three ways for himself, Tom and Dad. In addition, however, he left five "double-bar" opals with an expert cutter to cut on a special machine. These double-bar opals were nearly enough to put Jack's neck in a noose.

Dad was still in disgrace having first lost their opal find to thieves then wrongly pegging their claim. Their claim was now practically worthless. Tom was drinking heavily so Dad talked Jack into letting him go down the shaft to fill the buckets for Jack to haul to the surface. There, down in the hole, he died. Co-incidentally, just as Dad's body was lifted to the surface the over-zealous Police Trooper Burke rode up. He asked Jack if he had been alone on the field with Dad and then he questioned Jack about the sale of the opals. Jack forgot to tell him about the five double-bar opals he had left for cutting.

Even before the Trooper Burke found out about the stones, he was investigating the shaft, the windlass and the bucket for evidence that Dad had been murdered. But on being assured by the doctor that Dad had died from heart disease, he (reluctantly) allowed Dad's funeral to go ahead.

For Jack, Dad's death had taken the shine off Lightning Ridge. He told Tom he was leaving but then Trooper Burke found out about the five double-bar opals. When Jack returned to the Ridge from a brief holiday in Sydney, he was served with an exhumation application to be heard by the visiting Magistrate. Trooper Burke was still convinced that Dad did not die of natural causes and even more convinced that Jack was somehow responsible for Dad's death. However, he could not persuade the Magistrate so Jack walked out of the court a free man. Jack worked for a while longer at the Ridge but only to get enough money to start heading north – to the Cape York Peninsula.

His new life took him to Cairns, then Cooktown, three years of war and (ultimately) to Jack strolling with George across a coral reef on Howick Island.

13

Improvising
September 1920

Jack and George, on their Day Three exploration of their island, continued talking as they walked the kilometre or so across the reef at low tide to get to the Mound. They found the Mound to be even less promising than the other two hills on Howick. It was just a jumble of piled up boulders with almost no vegetation except for a few stunted bushes on the mangrove forest side away from the sea. As the last of the tide receded they sat on top of the Mound and looked out past Coquet Island to the peak of Lizard Island in the far distance.

Looking back across the reef George and Jack could see the two little hills that seemed to be growing straight up out of the mangroves. Suddenly, George jumped up, his face white, and explained that he had forgotten to bring his 'scope (as he called it) – the essential medical equipment he had left behind in Cooktown.

That George had left his vital medical supplies back in Cooktown was a disaster. Later Jack would say George should have stayed living near hospitals and doctors. He was probably quite correct but George was a bushman and well used to making the best of any circumstances. Even so, things must have looked bleak on that day. George had to tell Jack that he might die.[1]

George said nearly three metres of his bowel had been removed and he showed Jack "a fearful scar" on his abdomen beginning with a "*gruesome looking hole, with the flesh bunched around it.*"[2]

In the first book Jack said George's operation had been carried out in

Egypt. In the second *Madman's Island*, Jack said the operation was done in France. In fact the operation on George was performed in England.[3] Jack's mistake was excusable. It seems that George let Jack believe that his operation was the result of a war wound.[4] That was not true.

The "Great War" would have been no picnic for George, he was obviously seriously ill and the conditions of war in France were appalling. With hindsight he probably should not have been accepted for enlistment. He saw only three months of active service before he was admitted to hospital.[5] There were no heroics or war wounds. George had suffered chronic constipation and bowel trouble for years. This could easily have developed from living rough in the jungle and eating bush food. It is probably reasonable to speculate that this diet, plus possible alcohol dependence, finally killed him at the comparatively young age of fifty-four.[6]

George arrived in France in March 1918 but by June of that year he was in hospital and unfit for further service. From his record there does not seem to have been much done medically for George until he got back to England on leave. His condition was worsening and by November 1918 he was really sick. He was finally admitted to Devonport Military Hospital. Still nothing much happened. Even in hospital (as late as December 1918) George was still being reported as seriously ill with an intestinal obstruction. He then had the major operation that left the scars Jack saw on Howick Island.[7]

Now, although George's condition probably pre-existed his enlistment, something must have happened in France to exacerbate it. Back in Australia in September 1919, an army medical board found that George's condition (and subsequent operation) was caused through his active service. The Board also said George's previous condition was made worse by his Army service.[8]

In his Army medical records George had a laparotomy. That was a major operation that involved opening George's abdomen and removing his Ileum – the final section of his small bowel. He also had a colostomy where part of his large intestine was removed and part of his colon was connected to the wall of his abdomen leaving an opening to the outside of his body. This is what he showed Jack.

The "gruesome looking hole" described by Jack was the last step in the

procedure. The large intestine would have been drawn out through the hole and stitched to the skin. After the colostomy, George had to get rid of faeces through that hole. The irrigation procedure Jack describes would have been absolutely necessary.

Even today, one option for managing the faeces is by an irrigation procedure like that described by Jack. Seventy-odd years ago, however, the outcome of this operation would have been of great interest to George's doctors. Jack said he had heard gossip about George being asked to periodically inform doctors at Guy's Hospital that he was "still living".[9] His doctors would have been interested in how long he could survive and it could have been possible that they asked him to report every six months.

For the irrigation procedure today, the person places a catheter into the hole and flushes with water allowing faeces to come out into an irrigation sleeve. In the second *Madman's Island*, Jack wrote that George had told him he had a silver tube inside the hole but he was mistaken. In George's medical records there is mention of a belt (presumably to stop faeces coming out of the hole). The modern procedure uses a patch quite effectively.

While George was in hospital in London, he was shown how to irrigate his bowel and by the time he was discharged he was competently washing his bowels out every other day. He would not have been happy about this outcome but at least he was alive. For George to live through this experience was no small matter. For a start, his chances of surviving were decreasing when they did not treat him in France and did not immediately operate on him in London. The doctors seemed to have left his operation until the last minute. He was lucky to have survived.

At the end of Day Three when George suddenly remembered his irrigation equipment he was faced with the predicament of being on a deserted island with no hope of outside assistance for at least a month. He had only two choices – improvise or die. Nevertheless, George had an advantage over most people. George was a long-time bushman and men like him were masters of improvisation and self-medication.

Often in the past and far from any aid, George's life would have been threatened by illness or injury. He would have had to make do or die. It is interesting that after the initial shock of remembering that he did not have his equipment he dealt with the situation in a matter-of-fact way. There was no panic. Much to Jack's amazement he just carried on prospecting and did not start preparing his replacement equipment until that evening back at their camp.[10]

By firelight George set to work to make a tube and a funnel to replace the original versions given him by the doctors in London. George used some of their stock of fencing wire and some sacking from the strong hessian bags designed to carry tin. George cut a piece of fencing wire to the length he remembered. Then he bent it into shape according to his memory of the shape of the original funnel. He cut the hessian sacking into strips and pressed each strip around the wire model. It took him two hours by firelight to sew the canvas around the wire but he finished with a stiff and perfectly rounded canvas tube with a long curve in it. He cleaned an empty jam tin with sand and sea water then he hammered it flat and shaped it into a rough funnel.[11]

As soon as he had finished (and in the dark) George waded out to a flat rock just off the beach in the lagoon. On his side, he scooped up sea water in a billycan and poured it into the hole in his abdomen using his improvised funnel and tube. His idea was that the sea water would be a substitute for the chemicals left behind in Cooktown.[12]

His first model was not good enough. According to the second *Madman's Island* George had to lie on that rock all that first night. In the first book the procedure took two hours. This worked just enough to keep George alive but he still suffered great pain. Jack wrote that this pain was the cause of George's "madness" but this claim must be open to question. Not only did he live but George's improvisations allowed him to continue prospecting.[13]

Then, in both books Jack said they had a fortnight of bright clear days, gentle breezes and cool quiet nights. In this idyllic weather the two men thoroughly prospected the island. They found threads of quartz rich in tin and sometimes wolfram in the Peak but only in small patches. In the first book the two men were barely tolerating each other but in the second *Madman's Island*, for the two weeks leading up to their first month on the island, they worked together digging out the ore from the Peak.

Compared to the stormy relationship yet to develop, that fortnight was harmonious. At this time, in the second *Madman's Island*, Jack said George was a good resourceful mate.[14] During this period George and Jack must have talked about many things but for the first time (mentioned in both books) George expressed his wish to remain on the island. Jack was incredulous but George said, "*The world's got no time for me, and I've got no time for the world.*"[15] It was also at this time that George planted some sweet potatoes. The sweet potato cuttings or corms that George had brought from Cooktown were already sprouting and George had them in the ground some time during September. This date is important because Jack raided them in January (see Chapter Twenty-one). As Jack said, they take at least four months to be at all edible.[16]

One day, taking a break from their work, George and Jack were sitting on top of the Peak yarning.[17] At around this time, in the first book, Jack wrote that they discussed (not altogether amicably) George's ideas about living permanently on the island. In the second *Madman's Island* Jack wrote of a fortnight of bright days drifting by and one day taking a break from their work. Jack wrote that he and George talked about Captain James Cook who came so close to disaster in this area. As Jack said in another of his books, if it had not been for a plug of coral the *Endeavour* would certainly have sunk.[18]

Not far to the south of Howick Island everyone on board the *Endeavour* nearly died a lonely, unremarked death. Cook would never have been heard of again and the whole course of the history of Australia would have been different. As the two men sat there yarning in the brilliant sunshine they talked about Cook's Luck.[19]

14

Cook's Luck

Based on a brief mention in the second *Madman's Island*, George and Jack did talk about Cook and his near-disastrous adventures around Cooktown. They may have talked about the events grouped under the heading "Cook's Luck" because they are well known by people in Cooktown. This Chapter mainly refers to the journals of Captain Cook and Joseph Banks but many on board the *Endeavour* recognised history in the making. There were fourteen journals kept on the ship, two of them by civilians (Banks and Parkinson). Richard Orton (Cook's clerk) made three copies of Cook's original journal, one of which was sent home from Batavia in a Dutch ship. Cook's original handwritten journal was lost for over 140 years until it was purchased by the Australian Government in 1923 for £5000.[1]

When Cook and other European explorers were discovering the continent, Aboriginal people already lived here but the assertion that he declared the land to be "terra nullius" (or no man's land) so that he could claim it for Britain is a fiction.[2] Cook clearly recognised prior occupation and deliberately tried to avoid conflict with the "Indians" who lived here.

In discussing Cook's place in Australian history, many commentators give precedence to Dutch, French and Portuguese and even Chinese and Spanish explorers.[3] However, when, on 22 August 1770 at Possession Island in the Torres Strait Cook claimed the east coast of the continent for England he did, *in effect*, discover Australia. He had accurately charted most of the east coast and later named it New South Wales. Bowen and Bowen (2002) describe earlier voyages that left *Terra Australis incognita* a mystery for 200 years but they note, "*an increasing conviction among some Western historians is that the first European explorers to visit and chart the location of the reef were Portuguese*".[4]

Bowen and Bowen also describe the charts made by Spanish, Dutch and the French explorers but nevertheless, Cook discovered the country in the sense that other pioneers in many fields have made their own discoveries. In the field of prospecting, for example, many prospectors built on the work of others to finally and rightfully claim a discovery. One geographically relevant instance deserves a mention. William Hann found gold on the Palmer River but James Mulligan is awarded the credit of discovering the field that led to the creation of Cooktown. Mulligan is credited with exploring, assessing and then making public the extent of his discovery. This is exactly what Cook did with his discovery of the east coast – even though the country was already populated he discovered a land that Europeans only suspected.

History is full of people who found something and either did not recognise it for what it was or did not acknowledge the importance of their find. There is a point of view that a discoverer is not someone who happens to see something (minerals, for example, or land) and moves on. The discoverer is one who recognised, assessed and then announced. Blayney (2008) argued that other explorers left signposts for Cook (he might even have had some rough charts) but his work was more perceptive and thorough than that of any of his predecessors.[5]

Although other decisions ultimately led to the whole of the continent being colonised by Britain, Cook kicked off the process and every Australian should thank him for it. As Jack said, Australia would surely have been worse off if another nation had colonised the country.[6] Worse still, the continent would have been in turmoil if (back in Cook's time) a number of European nationalities had divided the continent into four or five countries (with four or five systems of law and governance). From the humblest of beginnings Cook rose to rightfully claim a place in the history of many countries, including a major role in the creation of Australia as the world's only country to occupy a whole continent.

All this was in jeopardy when the *Endeavour* struck a coral reef on 11 June 1770. But for an amazing run of luck, Cook and all his crew could have disappeared without trace. As Jack said – quite correctly – but for a lump

of coral the size of a man's fist the whole history of Australia could have been different.[7]

With today's infinite and immediate ability to communicate with anyone anywhere, it is hard to imagine a tiny wooden ship alone and isolated from any assistance stuck on a coral reef off an inhospitable coastline. But for sheer luck, all on board HM Bark *Endeavour* would have joined the long list of mariners who simply disappeared. They survived in North Queensland only through Cook's Luck. Cook was so lucky off the coast of North Queensland that he never again came near the place. In truth, "Cook's Luck" is an inadequate term to describe Cook's knife-edge struggle for survival over nine-and-a-half weeks in 1770. These weeks were the most hazardous and traumatic of the whole of his three-year voyage (including even the decimation of his crew after visiting Batavia, now Jakarta in Indonesia).[8]

Although he had no idea at the time, Cook's Luck really began in 1742 when he ran away to sea. At fourteen he got a job at Whitby aboard a collier. In those Whitby colliers Cook learned seamanship in the extremes of weather in the North Sea. More importantly, he learned to respect the design and seagoing qualities of these ships. This is where his luck really began because, decades later when Cook came to choose a ship to sail the world, he chose a Whitby collier.

Cook praised the former Whitby collier *Endeavour* for her seagoing qualities.[9] It was the basic design of the ship that allowed the *Endeavour* to survive the Great Barrier Reef coral. Whitby colliers were designed with a shallow draft, an enormously strong keel, thick planking and a flattish bottom. This construction firstly enabled the *Endeavour* to survive grounding on and then re-floating from a coral reef. Secondly, because of the coral damage, Cook had to beach the *Endeavour* to repair it. Cook's experience of Whitby colliers and his subsequent choice of ship allowed this manoeuvre. Almost any other design could not have been easily and safely beached.[10]

Some commentators have said Cook was following an earlier explorer's map.[11] Others have said Cook could not have had maps because he was still prepared to sail at night through all the many coral reefs along this coast.[12] Although Cook might have had one or more of these early maps, it is quite obvious that he did not know what was ahead of him. After nearly a year diligently mapping the east coast of the continent he was being funnelled into a maze of coral and rock. The further north Cook sailed the closer is the Reef to the coast.

Into this maze of coral Cook was sailing at night!

Cook and the Gentlemen (Banks' party) had just gone to bed when at 11.00pm the ship ran aground on what is now known as Endeavour Reef. Perhaps Cook was a little too confident of his luck.[13] Whatever the case (chart or no chart), in the middle of the night on 11 June 1770 the *Endeavour* was stuck fast on a coral reef with no hope of assistance.

Cook's Luck kicked in immediately. At this time of the year Cook should have encountered the powerful South-east winds that blow almost constantly. During this period there can be an occasional few days' respite from the wind; the grounding of the *Endeavour* occurred during one of these rare and short periods of calm. Cook wrote, "*Fortunately we had little wind, fine weather, and a smooth sea…*" (In this and subsequent quotations from Cook's journals, idiosyncratic spelling and out-of-date expressions have been minimally edited for ease of reading but the meaning of the quoted text remains true to the journals.)[14] Just a few days later the wind returned so strongly that it prevented Cook entering the river he later named the Endeavour River. Had this wind come up earlier the ship would have been ground to pieces.

Cook was also lucky that the *Endeavour* ran aground at high tide. The amount of water over the reef as well as the design of the Whitby collier minimised the damage. The high tide did, however, create its own set of problems. As the tide dropped, Cook's ship was stuck fast. Because Cook could not rely on tidal assistance alone to re-float the ship, he threw overboard some forty tons of equipment, ballast and the ship's drinking water. Included in the equipment thrown overboard were six cannons attached to buoys:

> "*… About the top of high water we… threw overboard our guns, iron and stone ballast, casks, hoop staves, oil jars and other material. All this time the ship made little or no water. At 11.00 am being high water as we thought, we tried to heave her off without success. The ship was not afloat by more than* (thirty centimetres) *notwithstanding by this time we had thrown overboard forty or fifty tonnes weight.*"[15]

After the *Endeavour* was re-floated Cook could not lift the cannons and abandoned them. For nearly two centuries many searches had been made for these cannons without success. There was even speculation that the *Endeavour* had struck some other reef. Then in 1969 an American team, using magnetic search equipment, found the cannons and an anchor Cook was forced to leave behind.[16] The anchor and one of the restored cannons are now in the Cooktown Museum.

Cook and his crew were in mortal peril. The *Endeavour* was in great danger of sinking. He could wait no longer because he could not assume the calm weather would continue. As the tide rose so the leaks in the hull increased. Nevertheless, Cook decided to risk all and float the ship off into deep water at the next high tide. Cook wrote:

> "*By this time it was 5.00 pm. The tide we observed now began to rise and the leak increased upon us. We set the 3rd Pump to work. We should have set the 4th to work also but we could not make it work. At 9.00 pm the ship righted and the leak gained upon the pumps considerably. This was an alarming and, I may say, terrible circumstance and threatened immediate destruction to us. However, I resolved to risk all. I decided to heave her off. Accordingly I turned as many hands to the capstan and windlass as could be spared from the pumps. At about 10.20 pm the ship floated, and we hove her into deep water. There was (*over a metre*) of water in the hold. This done I sent the longboat to take up the stream anchor. We got the anchor but lost the cable among the rocks. After this I turned all hands to the pumps because the leak was increasing upon us.*"[17]

Cook's Luck was holding! Not only did the weather remain calm but a piece of coral the size of a man's fist had broken off and stuck fast in the timber of the hull. If that piece of coral had not plugged the hole the *Endeavour* would have been lost. At that place the reef was steep. There was a depth of about twenty metres behind the ship. The *Endeavour* was taking on water.

All hands (even Cook and his officers) manned the pumps continuously. Just short of twenty-four hours after she struck the reef, the *Endeavour* was afloat again but the water in the hold was increasing. At this moment, a measurement mistake caused great fear among everyone on board. The mistake led them to believe the ship was sinking and some of the men gave up hope of saving the ship and prepared for the worst.[18] However, "*this mistake was no sooner cleared up than it acted upon every man like a charm. They redoubled their vigour, insomuch that before 8.00 am they gained considerably upon the leak.*"[19]

As the leak increased it was decided to draw a sail under the ship on the outside; a technique known as "fothering":

> "*Some hands were employed sewing oakham, wool and other substances into a lower steering sail to fother the ship... we mix oakham and wool together (but oakham alone would do). We chop it up small, and then stick it loosely by handfuls all over the sail. We then throw over it sheep dung or other filth. Horse dung for this purpose is the best. The sail thus prepared is hauled under the ship's bottom by ropes. If the place of the leak is uncertain, it must be hauled from one part of her bottom to another until one finds the place where it takes effect. While the sail is under the ship the oakham and other substances are washed off. Part of it is carried along with the water into the leak, and in part stops up the hole. This strategy re-inforced the coral plug so successfully that the crew were able to keep control of the leak with only one pum*p."[20]

Whether this fothering could be classed as more of Cook's Luck or good management of a dire crisis, Cook and his crew were now greatly heartened. Their hopes were raised for a chance of saving the ship but only if they could find a suitable harbour. Cook was considering breaking up the *Endeavour* to build another boat to carry them to the East Indies some 1500 miles away through the poorly charted Torres Strait:

> "*It is much easier to conceive than to describe the satisfaction felt by everybody on this occasion. But a few minutes before, our utmost wishes were to get hold of some place upon the mainland or an island to run the ship ashore, so that, out of her materials, we might build a vessel to carry us to the East Indies. No sooner were we made aware that the outward application to the ship's bottom had taken effect, than the field of every man's hopes enlarged. We now thought of nothing*

but ranging along shore in search of a harbour where we could repair the damage we had sustained."[21]

Now for two days Cook searched the coast for a place to beach the *Endeavour.* The ship was not in any condition to sail. In the ship's boats, the crew alternately towed the ship and ranged ahead looking for a harbour. Weary Bay was named for this arduous time but Cook's Luck was still with him.

Cook knew there was no suitable harbour behind him and in any case he would have been trying to sail against the wind. This would have been a near-to-impossible task for the *Endeavour* even if the ship had been undamaged. But the *Endeavour* was not capable of sailing. Cook had to press on but he had little idea what was ahead. Towed by the ship's boats, Cook laboriously continued north searching for a suitable place to beach and repair the ship.

On 13 June 1770 one of the ship's boats was scouting ahead and found the perfect place to beach the *Endeavour* – the river Cook named after his ship. This was an extraordinary element of Cook's Luck. In less than fifty kilometres he came across one of the only places on the coast where he could repair his ship and replenish the food and water he had jettisoned on Endeavour Reef. The sanctuary he found in the Endeavour River was better by far for beaching the ship than any other place Cook had seen on his whole voyage.

However, Cook was prevented from entering the river by four days of gale-force winds. This weather (characteristic of the season) would have meant total disaster if there had not been a break in the weather a couple of days earlier. Cook anchored a couple of kilometres offshore and took a small boat into the river mouth to lay buoys along the channel. On 17 June, the weather moderated and Cook decided to attempt sailing into the river. He ran the ship aground twice. On the first grounding they got her off without much trouble but the ship stuck fast on the second grounding.

Grounded on a sandbank, Cook had to contend with gale force winds and heavy showers of rain but he began getting the ship ready for repairs. He was not too worried about his latest grounding and on 18 June Cook had the ship warped into place using a gum tree against a steep beach on the south side of the harbour. The remains of the eucalypt tree are in the Cooktown Museum. The place where the tree stood and where the *Endeavour* was beached is now marked by a rock and a plaque.

Cook's Luck also continued to hold. He found that the Aboriginal people who lived on the river were friendly and helpful. His task of repairing the *Endeavour* and replenishing their supplies would have been much more difficult if he had needed to deal with hostility from the people who lived there. For just under seven weeks Cook and his party lived in relative harmony with the Guugu Yimithirr people whose descendants still live around the Endeavour River. It could even be argued that Cook's seven weeks' living on the banks of the Endeavour River represents the beginning of European settlement in Australia.

For several weeks the Guugu Yimithirr kept their distance. They were watching as Cook had the damage to the *Endeavour* surveyed and set carpenters to work. What they thought as they discussed these strange people, the ship, tents, guns and so on was (of course) not recorded but a positive relationship between the two groups steadily developed. While the ship was being repaired, the scientist Joseph Banks, naturalist Daniel Solander and artist Sydney Parkinson explored the area and met the local people. This was the first sustained contact between Aboriginal people and Europeans.

On 18 July (now a month after the *Endeavour's* arrival) Cook, Banks and Solander met five Aborigines that they had not seen before. The Aborigines approached Cook and his party without fear and later that same day several small groups of Guugu Yimithirr people visited the *Endeavour.* They were becoming more familiar with the crew.

Next day, however, the relationship soured over a misunderstanding about some turtles. Earlier the Guugu Yimithirr people had shared some fish with Cook's crew. On this day a group of them came on board and saw some captured turtles. Then followed a clash of cultures related to the concept of sharing. Jack had learned the difference when meeting Aboriginal people with his mates the Baird brothers and Dick Welsh.[22] For Aboriginal people the concept of sharing is all-encompassing: there is no need to ask, no need to give. The European notion of sharing that includes hospitality, generosity, gratitude and selflessness is dramatically different. Sharing for Europeans is a choice governed by circumstances. It is not a choice for Aboriginal people. George would have had to learn this difference when he was living with Aboriginal people on the Cape York Peninsula around Coen.

So naturally, the Guugu Yimithirr people thought they were entitled to a share of Cook's catch and began to take two of them. For his part, Cook was having a hard time feeding all the people on the *Endeavour* and he refused to share the turtles. Cook instead offered the Guugu Yimithirr some bread which they rejected with scorn.

The incident that followed this dispute was the low point of the contact between Cook's party and the Guugu Yimithirr people. The Aboriginals set fire to grass around the camp and caused some damage. In return Cook fired some bird-shot and injured one of the Aboriginal men. Despite this bit of hostility over a misunderstanding about sharing the turtles, the previously good relationship was soon restored:

> "*We got the second fire out before it got ahead but the first spread like wild fire in the woods and grass. Notwithstanding my firing (in which one must have been a little hurt because we saw a few drops of blood on some of the linen) they did not go far from us. We soon after heard their voices in the woods upon which Mr. Banks and I and three or four more went to look for them. We very soon met them coming toward us.*
>
> *As they each had four or five darts (*spears*) and not knowing their intention, we seized upon six or seven of the first spears we met with. This alarmed them so much that they all made off. We followed them for nearly (*a kilometre*) and then set down and called to them. They stopped also. After some little unintelligible conversation had passed they laid down their spears and came to us in a very friendly manner. We now returned the spears we had taken from them which reconciled everything... They all came along with us abreast of the ship, where they stayed a short time and then went away. Soon after they set the woods on fire about (*two kilometres*) from us.*"[23]

Banks recorded another lucky escape. Their store of gunpowder had been put ashore exactly where the fire blazed and (only days before) it had been stowed on board before the fire was lit.[24]

A few days later another congenial contact was made. A member of Cook's hunting party became lost in the bush. He met a group of Guugu Yimithirr people and sat with them. Not only did they not harm him but they also directed his way back to the ship.

So, for nearly seven weeks, contact between the two totally different peoples after beginning with suspicion, developed into familiarity and (with the exception of a misunderstanding) remained cordial. The Guugu Yimithirr people were extraordinarily tolerant of these strangers entering their country. It could so easily have been different. After all, these strangers were intruding on their territory and killing their game. Even when one of their men was (albeit slightly) wounded after the turtle incident, they still forgave Cook and his party.

Cook's own good intentions must also be taken into account. Throughout the contact Cook demonstrated respect and empathy for the Guugu Yimithirr people. Nevertheless, regardless of good intentions and forbearance from both parties it cannot be denied that Cook's Luck played a significant role in his ability to spend seven weeks, in peace, repairing the *Endeavour*.

Cook was anxious to leave. On their second day he climbed Grassy Hill – the Hill that towers above Cooktown. It is not grassy today because the Guugu Yimithirr people no longer burn off the undergrowth. It is today more difficult to climb but even so, just about everyone in Cooktown has stood where Cook tried to spy a way through the reefs outside the river mouth. Twelve days later he again climbed the Hill and his overall impression of the area was not improved.[25] He could see no way through the coral.

Cook stayed in the Endeavour River until 3 August. He was now being much more careful. He was not going to suffer another careless grounding in such a remote and dangerous seaway. As he proceeded north he sent boats out ahead to survey the course and take soundings. He was looking for a way out of all the coral. Cook anchored off Cape Bedford about 25 kilometres north of the Endeavour River and at Lookout Point about 32 kilometres east of Lizard Island and forty-five kilometres south of Howick.

While he sent a group off in the direction of Howick to investigate the Turtle Group of islands, Cook himself went to Lizard Island because he could see it was one of the higher islands. He aimed to find a way out through the outer reef to the open ocean. Cook had had enough of coral reefs so he eventually decided not to sail past Howick Island. From the Peak of Lizard Island using his telescope to scan the ocean toward Howick Island, the sight of what seemed to be an almost unbroken wall of rock and coral deterred him from discovering the inner passage to the tip of Cape York.

Instead he headed for the open sea and into even more danger. He surveyed the outer reef to the east and he chose Cook's Passage (south-east of Howick) to escape to the open ocean. Now all these years later modern mariners can discuss this mistake but with respect, to use modern knowledge to damn Cook's decision to escape the coral does him an injustice.[26] The *Endeavour* was just one small ship alone in more-or-less uncharted waters and impossibly far from any assistance. No-one else knew where they were. Cook had carefully surveyed the area south of Howick Island and then made a considered decision to head for the open ocean. This turned out to be an understandable but quite erroneous decision.

Cook passed Howick Island some 40 or 50 kilometres to the east but outside the reef. Then, just a bit further north, he nearly lost his ship yet again. The *Endeavour* was becalmed and huge waves were carrying the ship on to the outer Reef. The drama began in pitch-black dark just after four o'clock in the morning. There was no wind but all on the ship could hear the dreadful roaring of surf. Daybreak came and they could see vast foaming waves breaking on the Great Barrier Reef less than two kilometres away. With no wind to fill the sails the ship was quickly being carried to destruction on the reef. There was no hope of dropping an anchor because of the depth of the water. All aboard thought they were doomed.

Unable to sail clear, Cook tried towing the *Endeavour* with his small boats. Even then, one of them was not immediately serviceable. The idea of trying to tow a ship with two or three small boats is incredible. Cook had already done it just weeks earlier at Weary Bay but now the small boats had to contend with an ocean swell. How those sailors must have laboured! Slowly they turned the bow of the *Endeavour* to the north trying to slow the ship's progress toward the reef. By six o'clock they were less than one hundred metres away – only the breadth of one wave – from certain death on the reef.

The third boat, a pinnace, had been patched and was sent out to help tow the ship. Now there were just eighteen men rowing. The pinnace had three pairs of oars, a yawl two pairs and a longboat had just four pairs. Eighteen men were straining to tow a ship against an ocean swell.[27]

They had little hope of survival. At that point they were about fifty kilometres off the coast with huge waves crashing on to the reef between them and a most inhospitable shoreline. The three small boats were workboats, not lifeboats. If the ship had sunk the small boats would not have been able to save everyone and most would have been drowned. Banks estimated they got to within about thirty-six metres of the Reef. He wrote that their plight was truly desperate and they could only hope for a speedy death.[28] All Cook could do with the ship's boats was to try to align the ship's stern to the reef and hope for the best.

Then, without in any way denigrating Cook's superb seamanship, Cook's Luck came good again. At the very last moment there came a breeze so tiny that at other times it would not have been noticed. This helped the sailors labouring in the small boats to move the *Endeavour* away from the reef. Then the breeze died. They were still less than 200 metres off the reef.

The little breeze came back briefly but from their new position they could see an opening in the reef. Cook sent a boat to investigate. The skill and bravery of the men in that little boat is almost impossible to comprehend. The sailors were rowing in a huge ocean swell, deliberately approaching the mountainous surf crashing onto the reef. In this roaring, crashing, foaming and spray-filled maelstrom they found a tiny opening only as wide as the length of the *Endeavour*. But it was their only chance.

But to Cook's surprise when they approached the gap there was an ebb tide flowing out of it like a fast-flowing river. The ebb tide had the effect of pushing the *Endeavour* further off the reef. By noon the *Endeavour* was about three kilometres away from the coral but the danger was far from over. The ship was in a bay of coral and was again being driven by the swell back toward the reef.

Cook sent another small boat to check out another gap about a kilometre away while he and his crew continued the fight to keep the *Endeavour* from drifting toward the reef and disaster. On the strength of a favourable report about this opening in the reef, at two o'clock Cook decided to take the risk. He really had no choice but to try this narrow, dangerous gap. He was able to approach it because the tide had turned.

By sheer luck, (a) the gap in the reef was at exactly the right place (b) the slight breeze allowed some steerage and (c) the incoming tide allowed Cook to sail through the gap in the reef. On the tide rushing in through the gap the Endeavour surfed through Providence Passage to safety.

Just a few days before, after considering the route inside the reef past Howick, Cook had been pleased to leave the coral behind him and to head for the open sea. Now back inside the Reef and safe at anchor he expressed his joy:

> "*We were hurried through in a short time by a rapid tide like a mill race. This kept us from driving against either side even though the channel was not more than (*400 metres*)... It is but a few days ago that I rejoiced at having got without the Reef. Now that joy was nothing when compared to what I now felt at being safe at anchor within it.*"[29]

Cook recognised the extent of "Cook's Luck" and the dangers he had faced in North Queensland. He struggled home after sailing the globe (and found he had been given up for lost) in just one small ship. Never again would he go exploring with just one ship. It certainly was a struggle to get home. When Cook left the east coast of Australia he had lost only seven men but by the time he got home thirty-one had died from diseases contracted in Batavia (now Jakarta in Indonesia).[30]

George and Jack, sitting on top of the Peak, talked about Cook. They would have known many of the details of his voyage because Captain Cook is a revered figure with (at least European) residents of Cooktown.[31]

15

Visitors September 1920

George and Jack had been nearly a month on the island.[1] In the first book Jack wrote that he and George talked about their lack of success prospecting the island and they also talked at length about George's idea of living permanently on the island.[2] In the second *Madman's Island*, Jack said George was once again a good mate.[3] So, at the end of their first month on the island the relationship between the two men still seemed to be relatively harmonious.

Up to this time the issue of whether Moynahan's *Spray* would arrive was not discussed. Both George and Jack were working hard but for different reasons. Jack was working to get every last bit of tin bagged and ready for his departure from the island.[4] George wanted the tin bagged but he had no intention of leaving. He saw the bags of tin as the way in which he could get necessary supplies (like tobacco, tinned food and flour). So they worked.

During that first month they found quite a bit of tin and wolfram on the Peak but it was not in payable quantities. For the month's work they found about half a ton of tin and about a quarter of a ton of wolfram worth perhaps £100. Finding the tin was one task, mining it another and getting it into transportable form for sale was a different task altogether. They built a bush kiln to burn the stone.

On the beach near their camp George and Jack built a box of logs to a man's height and arranged so that a draught could blow into the bottom of the box. They put a layer of firewood in the middle of the box then tipped the tin-bearing ore in on top until the ore filled the box. Then they added more logs on top and lit the firewood. The idea was to build a slow, roasting-hot fire so that when it burned down they were left with a pile of whitish-red rock. While this rock was still hot they threw sea-water over it. Then it was easily crushed. They used a sieve to separate the grains of tin from the burned and crushed rock.[5] The evidence of this kiln can still be found on the island. Nothing will grow on the long-ago scorched patch of ground.[6]

One day, around midday, Jack was walking toward the well dreaming of the imminent return of Moynahan's *Spray* when he was amazed to meet a strange man on "his" island. Jack was embarrassed – acutely aware of his dishevelled appearance – when he looked at the clean, well-dressed visitor. Jack was conscious of his month's growth of beard and hair. Two other men stood looking down at them from halfway up the Peak.

There was no mention of these visitors in the first book but in the second *Madman's Island* book Jack writes about visitors from the light-house ship servicing the light on nearby Coquet Island. In his book Jack did not mention the name of the light-house ship but on his return to Cooktown Jack was reported as saying, "*During the whole time, we were there the captain of a cutter landed once, and a party from the survey ship* Fantome *landed for a few days.*"[7] The *Fantome* did indeed survey the Great Barrier Reef in the period that George and Jack were on Howick. When writing his book some eighteen years after the event Jack's memory might have played him tricks. Either he mistakenly called the survey ship a light-house ship or perhaps the *Fantome* did call at Coquet to survey (and not service) the light. Whatever the case, the *Fantome* returned to Cairns in November 1920.[8]

In that same report Jack mentioned a second visit – "*the captain of a cutter landed once*". This reference could have been about the sinister black-painted cutter of the opium smugglers that appeared off Howick in January 1921 (just before Jack left the island).[9]

Probably because Jack was expecting the *Spray* at any time, he did not ask the visitors to take him off the island. The visitors joined George and Jack at their camp but they would not accept a drink of tea. Instead George gave them two freshly caught fish and the visitors said they would return with some tobacco. The visitors rowed back to the small steamer anchored off

Coquet Island. On their return the visitors brought not only tobacco but also some jam, butter sugar and tea but (to George's later regret) no fishhooks. The skipper of the *Fantome* offered to take them off the island but Jack wrote that George laughed the offer away. Jack said he laughed but with much less mirth.[10]

So Jack and George were left alone as their second month was about to begin. So too were their troubles about to begin.

16

Sand flies
October 1920

All during their first month the weather had been kind. George and Jack had fine clear sunny days and a light breeze to keep them cool as they worked. On most days the swell of the sea on the reef was just a quiet background murmur and on just a few nights storms brought havoc from booming breakers and high winds. After these storms, in the calm of the next day, there were drifts of sand-discoloured foam piled up all along the shore.

The tin and wolfram they had found was all bagged and ready to go.[1] As well as starting his vegetable garden, George seemed to be spending a lot of time fishing. Indeed he seemed to have become obsessed with catching a big grouper that lived in a cave just offshore.[2] Eventually the big fish took all of George's fish-hooks.

George and Jack had been busy all through September finding, mining and processing the tin and wolfram. After a shaky start when Jack would not pull his weight in camp they seemed to have settled in to an easy sort of relationship. Not that they didn't have their differences.

Tension did seem to be building and their relationship was not helped by a silly incident at the end of September. As George and Jack were burning the ore George picked up a little slug of yellow metal. He yelled to Jack that they had found some gold. Jack promptly corrected him. Jack said the yellow was merely a copper stain. George, having spent decades making a living from prospecting gold was being corrected by a younger man. He was not happy. Jack said George never forgave him for correcting this mistake.[3]

George would also have been a bit put out by the fact that Jack ridiculed George's idea of staying on the island. Jack showed little interest in George's efforts at catching fish. Starting a garden interested Jack not at all. However, the incident of the copper-stained tin could easily have been a turning point in the souring of their relationship. Then early in October things quickly went from bad to worse. It started overnight.

At about ten o'clock one night around the first day in October George was lying in his bunk awake and thinking when he gradually became aware that the sound of the sea on the reef had stopped. Now that was strange. The sound of the sea had always been with them. It surrounded them in more ways than the obvious fact that they were on an island. It was like a living, breathing presence. Now there was just silence.[4]

He felt a slight sense of unease like a feeling that something was about to happen. He got out of bed and stepped out of the tent. The clear sky was alight with stars and a half-moon but there was not even a puff of wind. Everything was still and quiet. It actually was quite beautiful and peaceful standing there in the warm blue-black dark but the lives of the two men were about to become almost unbearable.

Jack was still asleep inside the tent when George was attacked by thousands – millions – of biting insects. George gathered some smoking wood from the edge of their fire and retreated inside the tent. He had just closed the flap of the tent and was building a smoky fire when Jack woke. He was being bitten by the sand-flies that followed George into the tent. The smouldering chocolate-brown drift-wood controlled the insects but filled the tent with smoke. They had no choice. It was suffer the smoke or the insects that they could hear outside. This experience (as described by Jack in both of his books) has been questioned by at least one commentator:

> *"Are the Northern sand-flies different from their Central brethren? I have had quite a number of painful experiences, over protracted periods, with sand-flies and they always ceased their labours promptly at dusk, coming to life again at daylight. It was only possible for horses to feed at night."*[5]

As Jack described the crisis, it was fortunate that their camp fire was still alight and that the dark wood was smouldering. Some softer types of wood would burst into flame and burned away quickly. Other wood like the chocolate-coloured drift-wood burned more slowly. It was this wood that George grabbed form the edge of the camp-fire. As the wood began to fill the tent with smoke Jack was up and awake, futilely cursing and slapping at the insects getting at his blood.[6]

With the tent filled with suffocating smoke, the insects stopped biting. George and Jack sat up all night choking and talking about their dilemma. They were trapped inside their little tent. They could hear the shrilling of millions of insects outside. They hoped the insects would be gone by morning but they knew it was not likely. What they did not know was that this night was just the start of four weeks of agony topped off (in the final week of October) when the two men nearly died of thirst!

This is where their tempers really began to fray. There was a dead calm sea and no wind at all. And toward the end of the third week they ran out of tobacco. Or, to be precise, Jack had used all his tobacco. George had been carefully preserving his share by rationing himself and also mixing his tobacco with dried seaweed and tea-leaves.

Inside the smoke-filled tent George and Jack were safe but they could not stay there. They had to eat. They had flour but they ran out of tinned meat. George had been fishing constantly to supplement their tinned food but now fishing was going to be essential for survival – sand-flies or no sand-flies. But venturing out of the tent exposed them immediately to clouds of vicious biting insects.

At that time Jack took to sitting for hours every day on top of the Peak. He would cover his face and arms in cloth and sit there in sweating misery. In neither of his books did Jack tell us what George was doing but it is not likely he was (like Jack) sitting for hours wallowing in self-pity. George would not have been idle. It was probably George who kept the fire going and did the cooking.

To go outside the tent needed careful preparation. Luckily their long trousers were still in reasonable condition so they bound rags around the top their boots and bottom of their trousers.[8] The two men wound cloth around their heads and plastered mud on exposed areas of skin but even with these precautions the insects could not be denied. After twenty minutes any exposed area, especially hands and wrists were red, swollen and painfully burning.

George realised that their only hope of survival was to catch fish but he had lost all his fish-hooks. He decided to make a fish spear. It is likely that their relationship would have been tense because their attitudes toward dealing with the sand-fly crisis were different. Jack sulked. George adapted.

While George made his fish-spear Jack just looked on "moodily".[9] George did not find making a spear to be difficult. He had learned the skill from living with Aboriginal people. George began with the fencing wire they used for pot-hooks and many other jobs. Using the face of an axe as an anvil he tapped the wire straight with a hammer. Then using a length of bamboo as a spear shaft he bound the wire prongs on to the shaft with fishing line.

When the spear was ready and George had taken off his boots so he could wade into the mangrove forest, he asked Jack to go with him. All the time George had been making the spear, Jack sat on his bunk watching wordlessly. Several times during the process George had to race outside to get wire, the axe and hammer and the bamboo. All the time Jack just watched, even as George came back with his face covered with biting insects. In his books Jack gives no hint that he could have joined in the spear-making. It would have been better if each had a spear to increase their chances of quickly catching fish. But it was not to be. Jack just sat watching. George was not happy when Jack flatly refused to go fishing.[10]

There is no way of knowing what Jack was thinking while George was making the spear or what he thought they were going to eat if they did not catch fish. Surprisingly, however, Jack was "incredulous" at George's suggestion. He argued that George would not survive the insects.[11] And he did not believe George would catch any fish.[12] However, George knew there were fish that lived permanently in the mangrove pools and that they hid in the roots of the mangroves at low tide. Other fish came into the mangrove creeks from the ocean to feed. Some of these were trapped as the tide ran but between George and the fish were millions of sand-flies.[13]

So George left Jack sheltering in the tent and stepped out. Immediately he was attacked. It would have taken all his will-power to keep going. There would have been so many insects on his face that brushing them away would have been useless. He struggled on with sand-flies biting at his lips and eyes and ears. Although he knew it was hopeless, he would still have tried brushing them away. Even as he waved his hands in front of his face his hands and wrists were burning intolerably.[14]

George plodded on toward the mangroves through the thick cloud of shrilling insects. The weather did not help. The sea was an oily calm with only tiny waves lapping the beach. There was no wind at all. He would have thought of Jack back in the hot suffocating tent. Jack was suffering but at least he was free of the insects.

Deep in the gloom of the mangroves George crept carefully across the mangrove roots searching for a tell-tale ripple or the movement of a fin. He was working through the close packed roots toward the more open water of a mangrove creek. As the tide runs these creeks become raging torrents filled with big fish and even sharks but now the tide was out and he was hoping to find some fish trapped in the pools.

Several times along the way he tried to spear a fish but there were too many roots and he was not quick enough for fish well-used to fighting for their lives. When George got to one of these creeks the going was easier. Instead of climbing across close-packed roots, he was able to walk along a sandy bottom with the shade of mangrove foliage above him like a canopy. It was like walking through a green tunnel through pools of water up to a metre deep. There were plenty of fish but spearing them was not so easy. They were alert and very fast.[15]

After a while George got lucky (or was it a cod that was unlucky?) when he crept up on a pool and saw the flutter of a fin just like a leaf moving on the sandy bottom near a root. Mimicking his Aboriginal friends, George froze above the pool with spear poised. This was not easy. The sand-flies were tormenting him but hunger and the hunt for food were imperative. Then the big cod moved just slightly away from the shelter of the mangrove root and he struck. There followed a fierce struggle between fish and man. The fragile spear was pinning the six kilogram fish to the sandy bottom. The fish was writhing and fighting as George leaned on the spear shaft until he could grab its gills. Success! George speared another fish before he walked back to the camp.[16]

When George arrived back at the camp he was a mess. His wrists, ankles and eyelids were red from sand-fly bites. But he had two good fish. Jack hurried out to throw some wood on the fire. George cleaned the fish. While they waited for the fire to die down to cooking-coals they retired to the safety of the tent. Jack asked George about catching the fish. George told him where he had speared them but then he proceeded to offend Jack by bluntly telling him that in future Jack could go and catch his own fish.[17]

On reading the two *Madman's Island* books, some people might view George's attitude as too harsh on Jack.[18] But after a month by themselves George and Jack were not getting along all that well and there is more than one way of looking at the same situation. Jack was only just doing his share around the camp; never doing more than he had to do. George had not forgiven him for gloating over his mistaking the copper-stained tin for gold. Now George came back to camp to find Jack sitting in the comparative comfort of the tent and expecting George to suffer the sand-fly plague to catch fish for him. George might perhaps be excused for being grumpy.

Early next day Jack set about making his own fish spear. This was after breakfast (or rather, after George had breakfasted on the remains of yesterday's fish). George did not offer Jack any of the left-over fish. He was going to have to catch his own breakfast. Jack had never made a fish spear before and it was not much good. In both the *Madman's Island* books Jack wrote about his attempts to spear fish.

He was not successful because he did not have the experience. Of course he was constantly attacked by the vicious biting insects. Even though he was learning about spearing fish as he went along, he worked all that morning until the tide started rushing up the mangrove creeks. He could not spear any fish at all. In the first book Jack's first unsuccessful attempt at spearing fish was just the one morning but in the second *Madman's Island*, he went out on two mornings.[19]

After his unsuccessful fish-spearing efforts Jack came back to camp very angry and badly bitten by sand-flies. George was lunching off another big fish he had caught while Jack was away. In angry silence, Jack prepared, cooked and ate a flour-and-water johnny cake. A johnny-cake is an early American term for simple unleavened bread more commonly in Australia called damper because the camp-fire is damped down to coals to cook it. The simplest version is flour, water and salt but more elaborate recipes are possible if materials are available. Some travellers carried raisins to add to the mix. But it was a poor meal indeed compared to George's fish. George did not offer him any fish but neither did Jack ask for any.[20] At this point the relationship between the two men was (to put it mildly) quite strained.

Again George had speared just one fish and that was just for him. That might seem churlish. George could have shown Jack how to make a spear and how to catch fish. Modern readers sitting in comfort while reading Jack's books could find it easy to see how differently George could have acted. But it was not all George's fault.

After Jack had eaten his johnny cake lunch he made himself another stronger spear ready for the next day. There could be no more fishing that day. The tide would not be out until after dark. Just on sundown, Jack left the tent with his spear. In both his books he said he was so hungry that he decided to brave the sand-flies and try to find something to eat along the edge of the reef.[21] Luckily for him he found a colony of big blue crabs and speared four. That was good news for Jack but ironically, the fact that he had caught his share of the food even further soured their relationship.

In the first book Jack ran laughing back to the camp; perhaps directing his glee at George. In the second *Madman's Island* Jack offered George a share and then laughed when George declined his offer.[22] In either case, however, Jack's gloating behaviour was fuel to the fire growing under their strained relationship. It got worse.

The three weeks of almost unbearable torment were nearly past them when George decided to correct another issue that was sticking in his neck. The two men could only survive in the tent by keeping the chocolate-brown drift-wood just smouldering away. According to Jack's writing of the story, during the night it needed constant attention or it would burn out and the insects would attack. To his credit, in his book Jack admitted that it was

George who made sure their nights were bearable. Jack hardly woke at all while George stoked the fire and sometimes had to get more wood.[23]

Obviously with the lesson of the fish in mind, George decided to get Jack to do his share of fire-tending. Perhaps he took the lesson too far. During the previous evening the two men had been talking reasonably amiably. With hindsight (and without living through the privations of life on Howick) George should probably have just asked Jack to do his share of maintaining the fire. Instead he arranged the smoky fire so that he got the smoke and Jack did not. This meant that Jack awoke to the biting of insects and he had to rush outside in the middle of the night to try to find some suitable wood. In the dark he was not able to find the chocolate-wood so he burned green leaves that filled the tent with acrid smoke. It was a miserable night for both of them.[24]

So for the first three weeks in October George and Jack endured the dead-calm weather and the sand-flies. Jack was not happy with George. George had reason to be unhappy with Jack but at least there was no open conflict. They fished separate areas of the island. Jack focused on the crab colony but during October he also learned to spear fish. Between them, they seemed to be living quite well. Jack wrote, "*Always afterwards, the crabs were our meat; the fish were fish.*"[25] Even so, they could not get used to sand-flies. For three weeks they spent as much time as possible in the tent; venturing out only to catch and cook their food.

In the first book there was no mention of the two men talking during this time but in the second *Madman's Island*, Jack wrote that they talked on at least one occasion. They talked about plagues of insects and also discussed the effect of the calm on Moynahan's overdue arrival. Obviously, in this dead-calm weather Moynahan's ketch would be sailing nowhere. If he did come he would have to be using his motor and consuming precious fuel.

Those three weeks of oppressive heat, no breeze and hiding in a smoky tent from the sand-flies seemed interminable. George and Jack thought nothing could be worse. They were wrong.

17

King tide October 1920

Seven weeks had gone by – with three weeks of sand-fly-induced torture – then there was a week of King tides.[1] The term King tide is used to distinguish between the normal monthly Spring tides. A King tide is much higher and only occurs twice a year when the sun and moon line up and when the earth's orbit takes it closest to the sun. Because the tide described by Jack in his books occurred late in October (or at the latest early in November) it is not entirely clear that the tide was really a King tide.[2]

King tide or not, about the middle of October George and Jack started noticing the tides rising higher. Fishing was becoming difficult because the mangrove creeks filled and stayed too full for them to spear fish. At the same time there were more crabs as they came out of the deeper water in the mangroves to feed.

The sea had now reclaimed almost the entire island except for their tiny campsite. All over the island only the tops of the mangroves were above water. Then came the day of the highest tide. George and Jack awoke to find that at the base of the Hill and the Peak even the grass was covered with sea water. There was still no breeze. In the cloudless sky the sun blazed mercilessly. There was total silence. Just when they thought their misery could get no worse they found the King tide had overflowed the well. While the heat, the smoky tent and the biting insects were bad, the loss of fresh water created a crisis that was genuinely life-threatening. The water in the well was salty and undrinkable. They had not thought to keep a reserve of water.

Jack thought they should try digging for water but they decided that digging in the extreme heat would only more quickly dehydrate them. They could only lie in the tent and try to wait it out. The next day the water receded but the grass and the well were still covered with sea-water. Jack tried signalling a steamer by setting the grass on the Hill alight. The signal was ineffective because Aboriginal visitors used to burn off the islands and the steamer would take no notice of smoke.

In the morning they inspected the well and decided to try bailing if the tide had dropped enough by the next low tide. They had no other choice. If bailing did not work tomorrow would be their last day. All day George and Jack waited for the tide to drop. That evening they bailed out the brackish water and dug down deep into the sandy bottom of the well. The water was still too salty. All night they lay awake in despair. There was no water and still the sand-flies swarmed in the hot, windless air.

Next morning – their third without a drink – they walked wordlessly down the sodden track to the well. With almost the last of their strength they bailed it out. Although the sea-water, at low tide, was clear of the lip of the well it was still salty. They hobbled back to the shelter of the tent.

Around midday Jack heard the little yellow birds chirping. The two men had hardly heard the birds since the sand-flies arrived and when the well overflowed, the birds were completely silent. Now they were chirping. Jack ran down to the well and George staggered along after him. The little birds were sitting around the stones, chirping and revitalised. Even at high tide the sea-water was not now polluting the well. The water in the well was a bit salty but drinkable. It was a narrow escape for the birds as well as for George and Jack. If they hadn't used the last of their strength to bail the well at low tide the water would still have been much too salty to drink. They would have died that day.

They lay on their bellies at the well and plunged their faces into the slightly salty water; drinking straight from the well like a pair of animals. They had survived but what a miserable month! They were out of tobacco and they had no more tinned meat but surviving together had improved their relationship.

For a few days George and Jack were talking again but you would never guess it from Jack's first book. The only mention of any discussion (also referred to in the second *Madman's Island*) was when Jack offered George assistance when he was obviously in pain. The near-fatal lack of fresh water probably had an adverse effect on George's bowels. Jack wrote that George glared at him with half-sunken eyes staring from a ghastly face and told him, "*to go to hell*".[3]

As the days progressed toward the end of October the well water cleared completely. George and Jack had their good fresh water back again. The sand-flies were still fierce but they were alive. In the first book Jack did not often write about the times that he and George were on good terms.

The second *Madman's Island* was quite different; in that book Jack said their narrow escape had made them good cobbers again for a few days.[4] Even so, Jack gave his readers no indication of whatever it was that he and George discussed.

For sure, at some time on the island they would have shared their knowledge and experience of tin mining. Perhaps it was on this occasion, when for a short time they were again talking together, that Jack talked to George about the adventures of which he wrote in *Back o' Cairns*.[5]

18

Cairns

George had been living on the Cape York Peninsula since about 1892 (when he was about 18 years of age[1]) but almost nothing is known of his early twenties. However, in 1908 he declared himself to be a miner at Truxillo. At that time the Truxillo mine was a bismuth, tin and tungsten mine on the banks of the Black Spring Creek about 250 kilometres in a westerly direction from Innisfail.[2]

It was in that year, 1908, that Jack made the break from Sydney and took his wandering path to Lightning Ridge. After the two occasions that he tried mining for opal[3], Jack had finally had enough of Lightning Ridge. By March 1912, at 22 years old, he decided to head for Queensland.[4] Years earlier he had dreamed of crossing the Queensland border[5] so now he was working his way north.

He arrived in Cairns around May of the same year but Jack wrote little about the city. Instead he travelled west to the town of Nigger Creek on the famous Wild River tin field. Nigger Creek in the Herberton district was said to have got its name after an Aboriginal spearing of some miners.[6] As recently as 2003 the name was finally and officially changed back to the Aboriginal name for the area – Wondecla.[7] Jack said the name means "Meeting of the Waters".[8]

In 1875, some thirty-seven years before Jack arrived, James Venture Mulligan (the same man Jack credited with discovering the renowned Palmer River gold field[9]) discovered tin ore in the Wild River near Herberton. Nothing much happened until 1880 when rumours began that Chinese miners were about to move in. A new exploration found one of the richest fields of lode tin in Australia. There were huge outcrops of black tin-bearing rocks. Mining began in April 1880 and by the end of May 100 tons of ore were stockpiled ready to be transported to Cairns.[10]

A rush started after this claim was registered and good finds of tin were found in the nearby Nigger Creek and other gullies where the Wild River had once flowed. The massive amounts of ore being won justified building on-site crushing plant that operated day and night. On one Good Friday 1000 tons of ore were crushed. From shallow mines, much of this tin was bagging tin – tin already marketable without further processing.[11] This was the type of mine in which Jack started work.

Jack's mine was up the hill away from the little tent township of Nigger Creek and the temptation of its two pubs. He was mining underground, alone in the dark with only a candle for company. Sinking a narrow vertical shaft down to bedrock then tunnelling horizontally along the long-dry bottom of an ancient stream.[12] Jack had learned this method of mining at Lightning Ridge where he was searching rock for seams of precious opal. Here he was searching the layer of heavy mineralised sand left along the bedrock following a thin line of black alluvial tin that he described as: "*little black match heads wedged in amongst the gravel*". Jack described squatting with legs crossed swinging his pick in the confined space. His only light was a flickering candle held in an iron pointed candle holder he called a "spider".[13] It is quite likely that Jack would have told George about the experiences he described in *Back o' Cairns.*

Even so, Jack was still writing. At the instigation of his campmates in Nigger Creek he even had a go at poetry. Jack said his first poem was written down in his mine by the light of a candle plugged into the wall. Jack said he was sitting in the long-dry bed of an ancient river writing his "pome" on a grubby piece of paper.[14] Jack recited his poem to his mates around the "Council Fire" of the romantic bush poet Old Mick. Jack wrote:

> "*That awful night came when I had to face the Council, and they'd made sure it was a full Council, too. Hardly daring to smirk at the expectant faces grinning round that bright fire, I mumbled through my first pome. I've got it beside me on the table in the old notes, but daren't insert it here.*"

Although Jack was not pleased with his first attempt at poetry, his mates said he should send it to "The Bullerteen"[15] Later in his book *Back o' Cairns*, Jack did include one of his poems, "*When Girlie Goes Looking for Nuts*"[16] but his career as a poet was short-lived. From that point (until he started writing books) he stuck to writing "pars" for the *Bulletin* and other publications. Writing aside, Jack stuck to his day job. Laboriously each man (sometimes alone or with a mate) hauled his hard-won wash-dirt up the shaft to the surface with a log windlass. Then they combined their energy to separate the tin from the soil.[17]

Collectively the individual miners hired a horse and dray to cart the wash-dirt to a community "sluice-box" by a creek. A sluice box is a long narrow wooden box with a slope down which water from a stream rushes. The water washed all the lighter useless material away and the heavier mineral grains gradually settled. As one man barrowed the wash-dirt down to the creek, another shovelled it into the head of the sluice-box. Other men separated the tin from the residue of heavy sands. One man with special expertise worked against the flow of water to continually shovel the grains of tin settling in the box up to the head again so concentrating the tin. In one day Jack and his mates put through the sluice-box all the wash-dirt it had taken a month to dig out.[18]

Finally in a smaller box the wash-dirt was streamed with a special shovel to separate the tin from the iron sand. The tin (now appearing as heavy black sand) was spread to dry on a canvas tent fly. Then it was bagged into little canvas tin bags each containing about fifty kilograms.

Even so, Jack welcomed the now more-or-less constant cheques he got back from the Sydney *Bulletin*. Mostly writing as "Gouger" but sometimes as "Up North" or another pen-name, he was now a regular contributor to the *Bulletin's* "Aboriginalities" page. These cheques were welcome because Jack was never very lucky at either prospecting or mining. The fact was that Jack got more out of his "*battered old schoolchildren's exercise-books closely packed with notes on Nigger Creek*" than he did out of the hole in the ground in which he toiled.[19] From those exercise-books (perhaps forty-five years later) he sat down to write *Back o' Cairns* – a book mainly full of stories about the people he met in and around Nigger Creek and Herberton. So with just enough money to pay for his boat fare and to live for about a month, Jack packed his swag and headed north to Cooktown.

19

First Fight
November 1920

At least for a short time after surviving the king tide overflowing and polluting their only water source, George and Jack were talking together. However, their brief period of friendship ended abruptly when – once again – Jack presumed too much on George's goodwill. Every trip out of the tent was a misery. The sand-flies were a continuous torment but since their experience with the well George and Jack had started keeping a supply of fresh water inside their tent.[1]

Someone had to keep these tins full, so, a few days after the well cleared (because they were out of water in the tent), there was George walking back from the well with two billycans of water. Toward him came Jack with a third billy and a pannikin. As they met on the track Jack dipped the pannikin into one of the billycans George was carrying. Jack had lifted the pannikin to his mouth to drink when George struck it from his hand. George then threw a punch at Jack's head. There is no argument about what happened (as described by Jack) but the real story is in why it happened.

George and Jack had now been on the island for two months. There was still no sign of Moynahan arriving in the *Spray* to pick them up. The tension between George and Jack had fluctuated but had nevertheless been building for eight weeks. After only a day or two on the island George had been exasperated by Jack's laziness around the camp but they seemed to reach a working arrangement.

In the first book this eight-week period was described by Jack as constantly tense.[2] Except for a few terse exchanges, Jack wrote of only two instances of prolonged conversation.[3] In the second *Madman's Island* there were more periods of prolonged conversation but Jack still described George as sombre, moody and sullen. Jack said he was consistently growling (or even snarling) his replies. Then as their troubles piled one on top of the other, the tension grew. Finally, in November they fought.

The first punch was thrown ostensibly over a pannikin of water but this is an over-simplistic summary. George was not well. Particularly at that time (probably because of dehydration) George was in pain. Jack said he was "*suffering terribly*" and wrote that when he heard George's "*quickly stifled groan... all the bottled-up anger went out of* (him) *immediately*".[4] The saltwater irrigation of his bowel was probably not really effective and the long period of fresh-water deprivation would have left George in a weakened condition. He would not have been able to carry the large kerosene-tin bucket full of water so he was ferrying water in billycans – two at a time.

Kerosene tins were about twenty-four centimetres square and about thirty-five centimetres tall. They were designed to carry about eighteen litres of Kerosene but with the top cut out and a wire handle attached to the top edge, a Kerosene tin made a handy camp container for water. A billycan or billy is a medium sized pot with a wire handle on the top edge used for boiling water over a campfire. A pannikin is a small metal cup.

Under vicious attack from sand-flies, George was hobbling back to the tent from the well. So far as he knew, Jack was sheltering inside the tent. Then Jack came strolling down the track not with the larger kerosene tin but with a billycan and a pannikin. This is a clear indication that he was not about refilling their store of fresh water. Jack was out to get some water for himself. When Jack walked up to George he dipped the pannikin into the water George was carrying. From Jack's point of view this saved him walking to the well. From George's point of view Jack was sponging off his work.

George's anger boiled over. He dashed the pannikin out of Jack's hand and yelled at him that he could get his own water. George swung a punch at Jack but it was feeble and Jack easily dodged. George stood in front of Jack trembling violently with rage and fatigue. He and Jack were face-to-face and screaming at each other.

Why did George react violently to Jack wanting a pannikin of water? For Jack the answer was simple. To him George was a madman to be compared with crazed patients he had seen during the war on a hospital ship.[5] George would have seen the situation differently. The incident could not be considered in isolation from all that led up to it. George's words as he yelled at Jack ("*Get your own water. Think I carry water for you, curse you*!") carry the feeling that had built up over two months of living with Jack on Howick Island.[6]

In both books this incident comprises Jack's first piece of evidence that George was mad. It was not evidence of "madness". To illustrate: if there had not been prior history of their worsening relationship (and if Jack had been doing his share of the work) Jack might have walked up to George and said something like, "Good on you, mate. I saw we were out of water. Can I have a cup? I'm on my way to fill the other billy". In that context George would not have objected to Jack taking a pannikin of water but that was not the context.

By that time George would have got to know Jack very well. His resentment at Jack's laziness would have been simmering and worse, he might have come to the realisation that Jack actually expected him to do the work. Eley alluded to this aspect of Jack's personality when she wrote: "... *the truth was he* (Jack) *couldn't quite identify with the pick-and-shovel brigade*"... and... "*he saw himself mostly as miscast among men who though good-hearted, were not his equal*".[7]

Rightly or wrongly, George had a go at Jack. In retrospect the incident appears trivial but back there on the island with the heat, the sand-flies and (only a day or so earlier) nearly dying of thirst it would not at all have seemed trivial. Jack thought George was going mad but without excusing George's violence, there must be some doubt about Jack's diagnosis.

Jack immediately turned and walked up to the top of the Peak. George would have been still trembling and weak. He might have gone back into the tent to escape the insects but when Jack came back down from the Peak (he still needed that drink of water) George was standing outside the tent. Then, abruptly, the sand-flies were gone.

At the very moment Jack came down from the Peak there came just the slightest breath of a breeze, something they had not seen for a month of dead-

calm weather. They felt the soft breath of breeze and George was watching the breeze rippling the grass. Jack raced back up to the top of the Peak as the breeze steadily freshened.[8] The wind blew away all the sand-flies. With the breeze steadily strengthening he strolled back down to talk to George. All thought of the previous altercation vanished with the sand-flies.

George and Jack climbed back up the Peak together and looked out over the mangrove forest with the welcome wind in their faces. By then the wind had become so strong that huge breakers were rolling in across the reef and the whole mangrove forest was rippling and swaying. Wind-blown clouds were streaming across the sky. In high good-humour they went fishing together.[9]

What a change. They had endured a month of calm, nearly silent weather and now the waves were again breaking on the reef and a strong wind had transformed their world. That night, for the first time in a month, the tent door was wide open and they did not have to put up with the smoky fire. For the next two weeks George and Jack passed the time companionably. They fished together and shared the tasks around camp. For long hours they talked into the night and during the days (at low tide) they walked around the island. They were almost in a holiday spirit.[10]

Now this picture is quite different to the picture painted by Jack in his first book. Writing of the two weeks after the sand-flies had gone, Jack fails to mention the good times they had together.[11] In the second *Madman's Island* they were not at each other's throats. Among other reasons for their improved relationship was that George was regaining his strength. The seawater irrigation of his bowel was working reasonably well and he had less pain. He just had to make sure that he carried out the irrigation procedure regularly – not just when he felt the gas building up.

On one particular day (after a good breakfast of yesterday's fish) the two men set off around the island. They talked for hours as they walked across the island until just after lunch when, as the tide came in, they sauntered back to camp. While they were walking they shared their knowledge about the woman well-known in Cooktown as the heroine of Lizard Island – Mary Watson.[12]

Living in Cooktown, both George and Charlie would have known about Mary Watson even if they could have had different points of view on this controversial story. The basic story of the tragedy of Mary Watson is quite simple. The story began on Lizard Island some forty kilometres in a south-easterly direction from Howick. Her fisherman husband had left Mary alone with two Chinese servants, Ah Sam and Ah Leong. She also had her infant son, Ferrier. The little group was attacked by Aboriginals who first murdered Ah Leong. Mrs Watson fought back and escaped in a makeshift boat. Mary and Ah Sam actually landed on Howick Island but did not find the well. All eventually died of thirst on Number Five of the Howick Group – now Watson Island. Jack could not get over the tragedy of Mary, her baby and Ah Sam drifting to this very island and not finding the water that might have kept them alive. The full story is not quite so simple. Nor is it free of controversy.

20

Mary Watson

In the second *Madman's Island* Jack said he and George talked about the Mary Watson story. The story has been the subject of, or mentioned in, many books. In their authoritative work, Falkiner & Oldfield (2000) include a bibliography but their book has little about Mary's early life.[1] Robertson (1981) wrote a reconstruction but her fanciful filling in of the gaps in Mary's diary affect the credibility of the work.[2] Nevertheless Robertson includes more detail about Mary's early life. The two books can be read together to get a more complete picture. Another small book published by the *Cairns Post* tells the basic story[3] but Jack's books treat the story only superficially.[4]

Mary Watson was descended from two reasonably well-off Cornish families. She had always wanted to be a teacher. Shortly after the family arrived in Australia, Mary – at only sixteen – left her family in Sydney. She first travelled to Brisbane and from there she landed a job in Cooktown as a governess.[5]

In Cooktown Mary met Captain Robert Watson who at forty-two, was the same age as Mary's father. He and his partner, Percy Fuller, were *beche de mer* fishermen based on Lizard Island. *Beche de mer* (also called Sea Slugs, Trepang or Sea Cucumbers) are marine animals related to sea-stars and sea-urchins. There are about 170 species found in Australia. Perhaps a dozen species are harvested from coral reefs in tropical waters. The main market is China but other Asian cultures regard the flesh as a delicacy and the internal organs as an ingredient for traditional medicines. The animals are sometimes eaten fresh (even raw in Japan) but Watson was catching and preparing the dried version.

Jack detailed the processing of *beche de mer* in his book *Coral Sea Calling*. He described the animal as resembling an ugly foot-long sausage. He named a number of different types of *beche de mer* including the highly prized "deep-sea black". Jack said they are boiled briefly in a cauldron with strips of bark to give them a reddish colour before being taken out and gutted. Then boiling continues for eight or ten hours. After boiling they are smoke-dried until, shrivelled and leathery, they are bagged and sold for export.[6]

Soon after their marriage on 30 May 1880, Mary and Captain Watson arrived on Lizard Island with twenty hens, ten ducks and two pigs. Their house had been constructed a year or so earlier by Watson and Fuller as a base for their fishing enterprise. Even today the remains of this cottage can still be seen on Lizard Island. It was about thirty-five metres from the beach and comprised two rooms and a smoke house made of large blocks of granite. It had an earth floor. It has been said there were other less substantial wooden huts (one of which, after Mary arrived, was said to be occupied by Percy Fuller).[7]

Mary must considered the potential for loneliness (she was to be isolated on an island with her husband away much of the time) because when she first went across to the island she took her twelve-years-old sister, Carrie, for company. For six months Carrie lived with her older sister but with the onset of the Wet season with its heat, humidity and storms, Carrie went home to her parents who were then living in Rockhampton.[8]

So, not only had the settlement been in existence for at least a year when Mary and Carrie arrived but the two women lived there for another six months. These two facts are significant when considering the reasons for the later attack by Aboriginals. For Mary there was no temptation to go home to her parents. The loneliness and hardships of living in a stone hut on an isolated island did not seem to bother her at all.[9]

The fascinating foundation of the story is the poignant diary written by Mary Watson. Still existing in the Queensland Museum, it tells a story of bravery and resourcefulness. It tells the story of a woman living with hardship and isolation and (later) facing death without even a trace of self-pity. All later accounts of the Mary Watson tragedy are based on her diaries. The most relevant diary entries are those that began on Lizard Island and later on the

voyage to Number Five of the Howick Group. Mary wrote less than 400 words yet after more than 120 years these few words still have strong emotive power. Because the diary entries are so terse, there are gaps in the narrative that later authors have filled with conjecture.[10]

Mary's diary records two trips back to Cooktown early during her stay on Lizard Island. It notes that when she got back to civilisation she actually missed life on the island. In her diary she makes it clear that she was anxious to return to Lizard Island.[11] In many respects her story is identical to that of pioneering women all over Australia. Mary's story is an example of the pioneering women heading out into the Australian bush to a life of isolation, loneliness and hardship. When she arrived on Lizard Island, Mary was only twenty years old and she had come to Australia from England just four years earlier. Mary's diary tells us that she just got on with her life. She was getting ready for Christmas and to preserve her Christmas pudding she added extra brandy to the mix.[12]

Mary had two Chinese servants with her. Ah Sam and Ah Leong were to tend the vegetable garden and to prepare the *beche de mer* caught by Watson and Fuller. The *beche de mer* were boiled in what had been a ship's iron water tank. The cubical tank had been cut in half to make a tub a bit over a metre square and not much more than half a metre deep.

When Mary became pregnant she stayed in Cooktown for the birth of her son. Thomas Ferrier Watson was born on 3 June 1881.[13]

Because Watson and Fuller had almost fished out the area around Lizard Island they decided they should search for new fishing grounds further north. They planned to sail to Night Island over 200 miles to the north. Anticipating a voyage of between six and eight weeks, they sailed on 1 September 1881. Mary and her baby were left on the island with Ah Leong and Ah Sam.[14]

So in true pioneering fashion, a young girl (only recently arrived from England and more recently married) was left on an isolated island alone except for two Chinese servants. The Wet season with its cyclones and heat was imminent and her husband would have known of conflict between Europeans and mainland Aborigines. Was Watson irresponsible in leaving

Mary on the island? That's a matter for conjecture but 120-odd years later it is hard (and perhaps unfair) to try to gauge Watson's thinking.

Three weeks passed without incident. Mary had either no idea that she was in danger or alternatively she was incautious. Mary should have been more alert to the signs of an Aboriginal presence. Two weeks earlier Mary had been told there were Aboriginal people on a nearby island. Only two days earlier, Ah Leong had seen smoke from the same direction. Still Mary did not feel threatened. Neither she nor her servants had made any preparations to defend themselves. There was almost no water in the hut and Ah Leong was away by himself working in the vegetable garden.[15]

The tragedy about to unfold would certainly have had a different outcome if Mary had been prepared for an attack. If the three people had laid in supplies, fortified the cottage, stayed together and used their armed strength in a concerted defence (or even a counter-attack) there is a strong possibility that they would have survived. As it transpired, the attackers were afforded the opportunity to ambush and murder Ah Leong. They also were able to ambush Ah Sam and lay siege to the cottage. Mary's lack of preparedness eliminated her support and she was left virtually defenceless.

On Thursday 29 September 1881, Ah Leong was missing. Mary's diary for 29 September 1881 reads: "*Blowing strong SE, although not so hard as yesterday. No eggs. Ah Leong killed by the blacks over at the farm. Ah Sam found his hat which is the only proof.*"[16] The most likely interpretation of this apparently cold-hearted report of Ah Leong's death is that over the past weeks of loneliness Mary had reduced her diary to a terse minimum and on the day Ah Leong died she recorded the events of the day as they occurred with a time-gap between each entry.

Some commentators have suggested Ah Leong's body was taken to the mainland and eaten. Inspector Fitzgerald on 9 November found dresses and infant's clothing at Murdoch Point. He said Aboriginal people at Murdoch Point admitted to killing and eating the Chinese man.[17]

Of course, after Ah Leong was murdered Mary was aware she was under attack. On Friday the Aborigines came closer to the cottage along the beach. She fired the rifle and revolver to disperse them. Because of the words "*fired off*' (rather than "*fired at*") her diary entries imply she fired warning shots: "*Friday September 30. Natives down on the beach at 7PM. Fired off the rifle and revolver and they went away.*"[18] However, on Saturday the situation worsened.

Ah Sam was also ambushed when he was going for water. Mary's diary for October 1 reads: "*Natives 4 speared Ah Sam 4 places in the right side and three on the shoulder. Got three spears from the natives. Saw 10 men altogether.*"[19] Now any man speared with even one Aboriginal war spear in his body would probably die. No-one could survive seven spears if they were war spears. It is far more likely that the spears were light fishing spears. Mary might have fired again so that the Aborigines again retreated and allowed time for Ah Sam to make the safety of the cottage.[20]

It was 1 October 1881 and Mary's husband was not due to return for about two more weeks. Her situation was desperate. Ah Leong was missing (probably dead) and Ah Sam had seven spear wounds. She had little food and even less water. Jack always had the highest regard for pioneering women and the hardships they faced. He said he could only imagine these terrifyingly lonely nights – a young mother besieged in a stone hut with little food and water, knowing there were murderous Aborigines patrolling outside.[21]

Over the years there has been a lot of talk about why the Aborigines attacked Mary. Most explanations centre on an unsubstantiated assertion that Lizard Island was a sacred initiation site for boys and the presence of women was an affront to the Aboriginals. There have been ridiculous attempts to excuse the unprovoked murder of Ah Leong, the ambush of Ah Sam and laying siege to a defenceless woman. In one of the more ridiculous versions, Robertson (2000) said that to cleanse a sacred site of a woman's presence, a "*graceful but ugly*" medicine man "*pointed a bone*" at "*his old victim*" and Mary fell to the floor writhing convulsively before fainting. It was this same medicine man (according to Robertson) who "*by means of invisible ropes, could climb up and down between earth and sky...*". [22]

In another version (postulated in the Captain Cook Museum, Cooktown), Ah Leong was said to have been armed. Polite Aborigines were only asking the intruders to leave their sacred island. In this version they only retaliated in self-defence. This version is a patently obvious whitewash of Ah Leong's murder. It also classifies Ah Sam's ambush and the subsequent siege as "skirmishes".[23]

Falkiner details and then dismisses another version of the reasons behind the attack on the Watson fishing base. In this version the Lizard Island attack was part of a wider military-type campaign by the Guugu Yimithirr people to reclaim their land. Of course this is romantic nonsense. There was no suggestion of a mainland uprising. This version tenuously connects Mary's attack with disparate raids on other isolated fishing stations.[24]

According to Falkiner and Oldfield, many years later an Aboriginal man claimed to be an eyewitness. William Daku said that, as a child of ten, he witnessed the attack. He said he saw the murder by spearing of Ah Leong. There was no suggestion of self-defence. Then he saw Ah Sam, carrying water buckets, also speared. He said Mary fired on the Aborigines but none were hit. While his people went to camp on a nearby island, Mary sailed away.[25]

Another group of supporters of the female-intruder-on-the-sacred-site theory was a recent group of documentary film-makers on Lizard Island. These people miraculously discovered evidence that nobody had ever before uncovered. They sent their discovery to "museum authorities" (perhaps authorities like those at the Queensland National Trust) that confirmed the find as being of religious significance and therefore "Mrs Watson had to go". However, Lucas (1968) pointed out that countless people had previously covered the same ground without ever seeing anything remarkable.[26]

Jack was inclined to accept (as the reason for the attack) the woman's presence on a sacred initiation site for boys but the Aborigines who attacked Mary did not just suddenly discover her on their sacred island. They would have known about the Watson/Fuller fishing base for some years and at the time of the attack, Mary had been on the island for sixteen months. She had been there for two Dry Seasons, the middle-of-the–year months which were the only time the Aboriginals could safely travel to the island. Indeed, for six months (during the whole of the previous Dry season) there had been two women on the island.

A second reason against the woman-on-a-sacred-island idea is because

only the Chinese men were attacked. If Mary had been the target of the Aborigines' moral outrage, they could easily have waited in ambush to kill her and not the men. She was unsuspecting. Instead the attackers murdered Ah Leong and speared Ah Sam from ambush, so relinquishing their advantage of surprise.

By far the most likely reason for the attack was simple opportunism. The Aborigines would have known about Watson/Fuller and when the small fishing party found Watson and Fuller were absent they took the opportunity to attack a supposedly defenceless woman and her two Chinese servants. But although unprepared, Mary was far from defenceless. She had her courage. She drove the marauding Aborigines away but she knew they would return. She knew she could not withstand a siege for another two weeks. She decided on escape. Mary decided to risk using the *beche de mer* tub as a boat. As well as the tub she took with her a small leaky punt. Perhaps it was in this vessel that she stored their supplies.

It has been suggested that Mary was allowed to escape or (as William Daku said) Mary left when the Aborigines went to a nearby island. In fact, Mary's escape was made possible only because her attackers went for re-inforcements. A large group of Aboriginals arrived while Mary was at sea. Within a period of about a fortnight after Mary left Lizard Island, the Watson/Fuller station had been ransacked and fires were burning all over the island. About forty of the re-inforcement party were seen with eight or ten canoes. Mary was right: the Aborigines were intending another attack. Obviously, the small fishing party who mounted the first opportunistic attack called for a war party to finish the job when they found that the Watson/Fuller station was virtually undefended.

To make her escape Mary gathered equipment, clothes and food and a new exercise book and pencil. She packed a saw, a hammer, her watch, jewellery and some money. She took a bonnet for herself and a pillow and an umbrella for Ferrier. The little water that was left in the cottage was poured into water bags. The voyage began on 2 October 1881.[27]

Mary's diary for October 2-4 says they left Lizard Island on Sunday 3 September but in fact it was 2 October and all the rest of her diary entries have the date wrong.[28] Her diary reads: "*Left Lizard Island Septr 3rd* (really

October 2nd 1881). *Sunday afternoon in the pot in which Beche-de-mer are boiled in. Got about three miles or four from the Lizards.*" (October 4th) "*Made for sandbank off the Lizards* (but) *could not reach it* (two lines illegible)."[29]

In one sense, Mary's story can stand alongside other great sea voyages of the area. By setting to sea in half of a ship's water tank, Mary escaped the spears of the Aborigines who had murdered Ah Leong and ambushed Ah Sam. Originally the tank had been a steel cube. Now, cut in half, Mary's vessel was only about 122 centimetres square and about sixty-one centimetres deep. Even then, the freeboard was reduced by the original hole in the water tank that created a semicircle some eighteen centimetres deep. It certainly was not seaworthy.

As Cook and many others found, these waters are hazardous even in a purpose-built and properly constructed boat. Mary's half of an iron cube had no sails and little means of steering. The tank must constantly have been on the point of capsizing. The only steerage was provided by the two mismatched wooden paddles that were used to stir the boiling bêche-de-mer. It is hard to imagine two people and a baby trying to cross some forty kilometres of ocean in a flat-bottomed tub only a bit over a metre square.

After escaping from Lizard Island on the afternoon of Sunday 2 October Mary, Ah Sam and Ferrier tried to land on a sandbank but could not make it. It is distressingly easy to imagine what Mary did not say in her diary. Her situation was dire. She was out at sea in blazing tropical heat in an iron tub. That tub must have heated to an unbearable degree. She had her baby and a rapidly diminishing water supply. With her was Ah Sam who had been speared seven times only a few days earlier.

The two people (and a baby) in that makeshift vessel must have had to sit quite still to avoid capsizing. They must also have had to bail continuously to prevent the tub from sinking. Mary wrote nothing in her diary about fear or despair while she suffered from sunburn and the dread of drowning with her baby.

Then they reached Howick Island. Mary's diary for October 5 and 6 reads:

"Remained on the reef all day on the lookout for a boat, but saw none." (October 6th) *"Very calm morning. Able to pull the tank up to an island with three small mountains on it. Ah Sam went ashore to try and get water, as ours was done. There were natives camped there so we were afraid to go far away. We had to wait return of tide, anchored under the mangroves. Got on the reef – very calm"*.[30]

Jack said it was a pity that Ah Sam was not a bushman. If he had been, Jack said, Ah Sam would have seen the birds that live around the well on the island and so discovered a source of fresh water.[31] It is certainly sad that the little party passed so close to a source of fresh water when they had used all their water. But Jack's assessment of Ah Sam was too harsh. Only five days earlier Ah Sam had been speared seven times! He certainly would have been suffering. Even to get ashore must have been an ordeal.

Ah Sam might or might not have actually seen Aborigines but he could hardly be blamed for jumping at shadows. And the well was not easy to find. Even today the well is very difficult to find. As it turned out Mary (with Ah Sam and Ferrier) left the relative safety of Howick and drifted onward. Next day they headed for Number Five of the Howick Group (Watson Island).[32] Mary's diary for October 7 reads: *"Made for an island four or five miles from the one spoken of yesterday. Ashore but could not find any water. Cooked some rice & clam-fish. Moderate S.E. breeze. Stayed here all night."*

Watson Island is about sixty kilometres in a north-westerly direction from Lizard Island. They landed on this desolate, waterless island. On October 7 Mary saw a steamer and tried to signal it but no-one saw her.[33] Mary's diary for October 7 continues: *"Saw a steamer bound North. Hoisted Ferrier's white and pink wrap but did not answer us."* Surely most people would have felt a sense of hopelessness or despondency but Mary's diary revealed none of these feelings. She, Ah Sam and Ferrier were stranded on a tiny coral island. They suffered extremes of day-time heat and night-time cold.

The wind was blowing so hard they could not again launch the tub until it was too late. What a horrible predicament. Mary knew her baby would die unless there was a last-minute rescue. But nowhere in her diary is there any indication of her losing heart nor is there any sense of self-pity. Mary's diary for October 8 reads:

> *"Changed the anchorage of the boat as the wind was freshening. Went down to a kind of little lake on the same island (this done last night). Remained here all day looking out for a boat; did not see any. Very cold night; blowing very hard. No water."*[34]

Obviously, even as they were progressively weaker from thirst, Mary and Ah Sam still had some thought that they might again have use for the tub but the wind was too strong to consider launching it. They were stuck on the island. Mary's diary for October 9 reads:

> *"Brought the tank ashore as far as possible with this morning tide. Made camp all day under the trees. Blowing very hard. (no water)."*[35]

Later that day Mary gave Ferrier a dip in the sea but she had no water to give her poor dying baby. She knew now that they would all die unless someone came to their rescue. They had no more strength to launch the tank. Mary's diary for October 9 continues: "*Gave Ferrier a dip in the sea – he is showing symptoms of thirst – and took a dip myself. Ah Sam and self very parched with thirst. Ferrier showing symptoms.*"[36]

On Sunday Mary thought she would have died during the night. Mary's diary for October 10 continues: "*Ferrier very bad with inflammation; very much alarmed. No fresh water and no more milk but condensed. Self very weak, really thought I would have died last night Sunday*".[37]

Mary's diary for October 11 reads:

> *"Still all alive. Ferrier much better this morning. Self feeling very weak. I think it will rain today, clouds very heavy. Wind not quite so high. No rain. Morning fine weather. Ah Sam gone away to die – have not seen him since 9. Ferrier more cheerful. Self not feeling at (all) well."*[38]

Then Mary Watson's last poignant diary entry reads: "*Have not seen any boat of any description. (No water. Near dead with thirst.)*".[39] On or about 11 October 1881 Mary Watson, Ah Sam and Ferrier were dead. Their remains would not be found for more than two months – until 18 January 1882. In October no-one yet knew they were even missing.

The tragedy was compounded by the fact that, soon after they all were dead, the rain Mary had seen coming actually did arrive. When Mary was found the tank was full of rain water.[40]

It was not until about a week after Mary, Ah Sam and Ferrier had died that it became known they were missing. A sea captain and friend of the Watsons sailed past Lizard Island. He saw numerous fires in the bush on the island. Getting closer to land he tried signalling the settlement but (of course) there was no answer. Two Aboriginals were near the hut and the door of the hut was open. He thought the station had been abandoned. He sailed on to Cooktown. By that time the Cooktown Courier said there was much anxiety in Cooktown for Mary.[41]

A day or so later some Chinese seamen reported the same bushfires and about forty Aboriginals and eight or ten canoes on the island. These were the re-inforcements summoned by Mary's attackers.

On 21 October Bartley Fahey, who was Harbourmaster and Sub-collector of Customs as well as Water Police Magistrate, conveyed these reports to the police. Inspector Hervey Fitzgerald agreed to send an investigation party with Fahey to Lizard Island. Lieutenant Izzat of the HMS Conflict offered to transport the party.[42] When the police arrived on Lizard Island they found the cottage empty and wrecked.[43] The Watson's personal possessions were scattered and there were Aboriginal tracks around the cottage. They found Ah Sam's blood stains inside the cottage. Tracking the Aborigines they found clothing at a deserted campsite. They took Watson's store of *beche de mer* and anything else of value back to Cooktown.[44]

In searches of mainland Aboriginal camps some of Mary's property was found. Searchers also found the hair of a Chinaman.[45] All this time Captain Watson was still fishing and had no idea of what had happened until he was found early in November at Restoration Island – many kilometres from where his wife and child lay dead. It was now nine weeks since he had sailed away from Lizard Island and his wife had died about three weeks earlier. Neither Watson nor anyone else yet knew Mary was dead. All of Australia was following the sensational news of the missing woman.[46]

During this search there were many conflicting reports about what had happened to Mary, Ferrier, Ah Sam and Ah Leong. Adding to the confusion were multiple confessions of guilt by Aboriginal people living over a ninety kilometre stretch of the mainland coast. Even before the bodies had been found some of the Aborigines who had confessed had been severely punished.[47]

No-one knows for sure what constituted severe punishment but William Daku, (the Aboriginal man from Elim Mission on Cape Bedford who said he was an eye-witness to the attack) said he saw many people shot by police – including women and children.[48] But the confessions were wrong and the severe punishment was premature and unjust. The bodies of Mary, Ferrier and Ah Sam would not be discovered for another seven weeks, dead of thirst and not spear wounds.

Then for nearly three months Watson and the police searched the mainland and islands between Cooktown and Cape Melville. Watson had briefly visited the island and noted the absence of the tub. He ignored arguments that the Aborigines would have broken up the tub for spear tips and set about finding it.

On Monday 7 November Fahey cabled his superiors that Watson had told him there had been a punt on the island. Not long after this report, many weeks before his wife's remains were found; Watson found the punt on Number Five of the Howick Group. He appears to have missed the tub hidden in the mangroves.

Over three months after the deaths of Mary, Ferrier and Ah Sam, the crew of a *beche de mer* boat anchored off Number Five of the Howick group.[49] On Thursday 19 January 1882, Captain Bremner on his schooner *Kate Kearney* found the bodies. Some of the crew had gone ashore and one, searching for bird's eggs, found the partially decomposed remains of Ah Sam. In the subsequent search the tank was found.

Because the tank was half-filled with water, the bottom of the tank was perforated with an axe and a rifle was fired through the sides.[50] These marks can still be seen on the tank as it exists today in the Queensland Museum. The captain and his crew found the fully clothed skeleton of Mary Watson with

the baby embraced in her right arm at her breast. They covered the tank with a sheet of iron then sailed to Cooktown to report their discovery.

Less than a week after the bodies were discovered an inquest was held on board a ship anchored off Number Five of the Howick Group. On 23 January 1881 Bartley Fahey, Water Police Magistrate (the same Mr. Fahey who investigated Mary's disappearance from Lizard Island) went with Inspector Hervey Fitzgerald to recover the bodies. With them was Captain Watson. On 24 January a coronial inquest was held on board the government schooner *Spitfire* anchored off Number Five of the Howick Group. The conclusion reached was that there were no suspicious circumstances – Mary, Ferrier and Ah Sam died from thirst and exposure on or about 12 October 1882.[51]

There was nation-wide interest in Mary's disappearance and then national grief at the news of her tragic death. The extent of this outpouring of grief had not been seen before in Australian history and only rarely ever since. Mary and Ferrier were given a very large funeral. Ah Sam was laid to rest in a separate Chinese funeral held directly after that of Mary and Ferrier.

Watson never returned to Lizard Island. In fact both he and Fuller gave up their occupations at sea. Watson took up mining and this occupation eventually killed him. He died at Cooktown Hospital in 1894 of miner's pthisis.[52]

People from all over Australia were touched by the sad story. Mary's bravery was recognised by a public subscription that in 1886 enabled the building of a monument in Cooktown. Today however, Mary Watson's story has been de-emphasised presumably because of the connection with Aboriginal issues. This was particularly so in Cooktown's James Cook Museum. Until relatively recent times the story was prominently displayed in the original James Cook Museum opened by Queen Elizabeth in 1970. At that time there was a substantial display that included a painted backdrop and a replica of Mary's tank. The words of her diary were displayed. A substantial tribute to the early Chinese included a specific mention of Ah Sam and Ah Leong.

By 2009 Mary's side of the story had been almost completely removed from the museum. At that time there was available only a postcard, a booklet and an information sheet. A distorted version of the story appeared on two wall posters that excused the role of Mary's attackers. A replica of Mary's tank (part of the original display) was dumped with some rubbish at the back of the museum. This re-written history trivialised a significant event in white/black relations in Australia and distorting the story denigrated the bravery of a pioneering woman defending herself from a murderous attack and then undertaking a sea voyage of astounding difficulty.

Of course George and Jack, talking about Mary as they strolled around their island could not know any of this still-to-be-rewritten history. As men of their time and living much closer to the event, George and Jack would have been far less sympathetic to the Aboriginal point of view. In their day the Aboriginal side of the story got the same treatment that modern commentators have given Mary's side. George and Jack would have known about the summary retribution applied to probably innocent Aboriginal people after Mary's disappearance but they would be amazed to find how far the pendulum has swung in today's version of the story. They would be incredulous if they could be told that Mary's bravery and fortitude are today discounted.

Revising historical events is always risky because social mores change. Risky because the realities of isolation, privation, violence and survival no longer apply in genteel circles like the National Trust of Queensland. At least modern commentators should strive for some balance. The Watson story has a place in the history of pioneering women. It has a place in the history of black/white relations. It has a place in the history of Australia's system of justice. Whatever the context, Mary Watson's story has an indisputable place in Australian history.

21

Wind
November 1920

It was in the last week of October 1920 that George and Jack strolled across Howick Island talking about Mary Watson. The dead calm weather had gone and the sand-flies had been blown away by what was then a pleasantly strong breeze. That breeze, though, had strengthened into a howling gale.

Of course George and Jack were delighted when the sand-flies were blown away after a month of misery. Their relationship improved and they talked companionably long into the night. Next morning, when they went fishing together, the wind had strengthened. Jack said a howling gale nearly carried them off their feet.[1]

For the first two weeks of November the gale persisted.[2] Moving about was difficult. To try to climb the Peak would have been too dangerous – the wind was just too strong. There was no point anyway. No boat would be sailing past their island in this weather. The sea was incredible. George and Jack watched lines of breaking surf come crashing on to the reef with an almost continuous thundering roar. All day and all night the air was filled with spume flying off the top of the waves.

At least the two men could still catch fish and crabs. They had plenty of fresh water and best of all, they were still talking. Jack did not include this period in the first book but although they soon tired of the wild wind, George and Jack were safe and relatively comfortable. Even though they had fought at the end of October, in the weeks that followed they were the best of mates.

During the first two weeks of November 1920[1] Jack wrote that he and George, "*...yarned on the friendliest of terms....*"[2] but he did not tell his readers what it was they yarned about. There can be no doubt that at some time they shared their knowledge of Cooktown. This tiny isolated town figured largely in the lives of both men. Both of them would have had good reason to remember Cooktown and the people who lived there.

22

Cooktown

It would have been about October 1912 when just a month past his twenty-third birthday, Jack arrived in Cooktown. George was there already. At the time Jack arrived, George was probably prospecting further north in the jungle around Coen. In 1913 George was a miner at Mt Windowie and at Spion Kop, Yarraden. He was still at Spion Kop in 1914.[3] George did not ever make his fortune as a gold miner but it was twice reported (in 1913 and in 1914) by the *Cairns Post* that he was winning a reasonable amount of gold.[4] However, George would have arrived in Cooktown much earlier than 1913 – he had probably arrived some ten years earlier – around 1892.[5]

Unfortunately, little is known about George's arrival in Cooktown but he did from time-to-time live at Cooktown's West Coast Hotel. When he enlisted for the First World War in 1917, George nominated the West Coast as his "permanent address"[6] and Jack said George was "honary" barman when he met him in 1920.[7] George seems to have based himself at the West Coast and worked for his keep whenever he was in Cooktown.

The West Coast Hotel was one of the first substantial hotels to be built in Cooktown. It survived fires and cyclones and Cooktown's roller-coaster economy to continue to trade on the same site today.[8] Mrs Devaney was the publican when Jack arrived in 1912. When George arrived at the hotel in the early 1890s the publican was William Sleep.[9]

According to Jack, George had spent many years as a solitary prospector in the far north of the Cape York Peninsula.[10] When George arrived, Cooktown would still have been a reasonably large town. Indeed, at 2630 people, the size of the town was not too different from today's population. However, during the time that George lived around Cooktown he saw most of the population leave because of cyclones and fire. By the time Jack arrived in 1912 only a few hundred people lived in the town.[11]

Although George's arrival in Cooktown can only be conjecture, Jack provided a good description of his own first impressions of the town. Jack arrived on the little mail steamer, the *Musgrave.* He wrote that just about everyone in town turned out to make the arrival of the *Musgrave* a real social occasion.[12]

Jack was standing at the rail of the *Musgrave* being given a running commentary of the little crowd by one of the *Musgrave's* deckhands. One of the characters pointed out to Jack as he looked down from the deck of the mail boat was Captain Dan Moynahan. Even at first sight Jack remembered the black-bearded Moynahan because he towered above everyone else in the crowd. In his ketch, the *Spray*, Moynahan ran supplies and passengers from Cooktown up and down the North Queensland coast.[13]

Much later, when Jack became a popular author, he often wrote about Moynahan. He had good reason to remember him. In 1914 Moynahan and the *Spray* delivered Jack and some mates on Cape Melville to do some prospecting.[14] When he came back to pick them up it was Moynahan who broke the news that the country was at war. And it was Moynahan who delivered George and Jack to Howick Island.[15]

When, from the deck of the *Musgrave*, Jack first saw Dan Moynahan, he was impressed by his imposing physique and great black beard but later he came to respect him as a man who had been out there and done things – a genuine man of adventure. By the time Jack arrived in Cooktown Moynahan had been building his own ships and sailing the treacherous waters of the far north Queensland coast for nearly thirty years.[16]

Jack did not get to meet Moynahan on that occasion but as he strolled up Charlotte Street from the wharf he met two people who immediately offered him assistance. Jack was feeling bad. He was humping his swag and he had only £6 in his pocket. As he trudged past the Seaview Hotel, laughter from

a row of young girls leaning over the balcony (might he thought) have been directed at him. Mooching along with no particular destination in mind and very little money to support him Jack's mood was low until the local barber stepped out of his shop to give Jack a friendly greeting.

George Walmsley, barber, bookmaker and sports store owner always had a story, an opinion and a joke. It was George who suggested Jack board at the West Coast Hotel further up Charlotte Street. In his friendly fashion, he suggested Jack should get a job until he could properly equip himself for the bush and it was George the barber who gave Jack his first encouragement to try gold prospecting.[17]

That's even the sort of welcome George Tritton might have got from another friendly local resident when he arrived in town twenty years earlier. It is even possible that on his own first day in Cooktown George might have been directed to the West Coast Hotel. The West Coast became George's constant but occasional home base for the rest of his life.[18]

So at George Walmsley's insistence, Jack wandered up toward the West Coast. A bit further up the street Jack said he met a man called "James Dickie".[19] Unfortunately, it is likely that in his notes (and referring to them forty-six years later) Jack confused two notable Cooktown identities – John Dickie and James Dick. There was no-one in Cooktown by the name of James Dickie. No doubt from Jack's notes, his biographer Beverley Eley repeated Jack's error.[20]

While both John Dickie and James Dick fitted Jack's description and both had been explorers and gold prospectors, John Dickie did not have a house and wife in Cooktown. Jack met James Dick who owned the Little Wonder Store.[21] Elsewhere, in his book *My Mate Dick*, Jack got the two names straight, indeed, he devoted three chapters to describing John Dickie.[22]

James Dick lived for over forty years on the Cape York Peninsula. He was not only a merchant and businessman. He was also a renowned prospector, explorer and miner but his interests and vision were far more extensive. This extraordinary man was also a writer, historian, agriculturalist and a community leader. To the end of his life he was a great advocate for Cooktown and the Cape York Peninsula.[23]

John Dickie was an even more extraordinary character. Known as the "Stringybark Fox" and renowned for discovering the Ebagoolah gold field, Dickie was someone about whom Jack could easily have written several books. His reputation as an explorer and prospector was without equal. For example, in 1887 Dickie was landed at Cape Weymouth, north of the Lockhart River and nearly at the tip of Cape York. Alone he walked back to civilisation, living off the land.[24]

Then in 1896 Dickie walked with three pack horses from Darlot Goldfield in Western Australia, via Alice Springs to Cloncurry in Queensland.[25] Dickie always had two pistols on his hips. He had one for personal protection and a longer barrelled pistol for game.[26] Rather than pistols, on the long ride from Western Australia, he used his wits to survive. Passing through a gap in a range he saw a large number of Aboriginals on both sides high above him. They were just keeping pace with him but he knew an ambush was imminent. Just on dark, when he halted for the night, he hung balls of phosphorus from trees and exploded some dynamite. This display of seemingly supernatural forces scared the Aborigines so much that in their haste to depart they left a large number of spears and other possessions.[27]

It is surprising that Jack could have confused these two remarkable men but so be it. It would have been James Dick, proprietor of the Little Wonder Store who invited Jack to his home and offered him supplies to keep him going until he made some money. Jack was impressed that two strangers had befriended the swagman who had just come to town. Too shy to take up James Dick's offer of hospitality, Jack walked on up to the West Coast Hotel (although Jack did not say so, it is an intriguing thought that George could even have been behind the bar on that day). However, Jack did meet the licensee, Mrs Devaney, her son Paddy and Mrs Devaney's two daughters. Again he found friendliness and hospitality.[28] Cooktown was looking decidedly better!

The West Coast Hotel was certainly the right place for Jack to start to know Cooktown and its characters. Tin-scratchers lounged on the benches outside the pub along with cattlemen and sandal-wood getters. Gold miners from Batavia and Coen, in town to blow their money, excited his interest in travelling north. During those first few days in Cooktown he met many people who became his friends over the coming years.[29] Jack stayed at the West Coast for a day or so at a shilling a meal and a shilling a bed.[30]

Then Jack trudged along dreaming of riches from sandalwood and Hughie Giblet who was known as the Sandalwood King (although in his notes, Jack seems to have mistakenly written the name as Gilbert[31]). As Jack (with no horse at all) thought of Giblet's huge teams of horses, he started walking the forty kilometres along the Palmer Road out of Cooktown, past Black Mountain to the Lion's Den Hotel at Helensvale.

He stayed overnight at the Lion's Den and the licensee, Mrs Watkins, provided him with the best bed he had had for years. Next day he continued walking to the end of the road – to Rossville. There he took on work as a labourer for the Annan River Tin Mining Company.[32] Jack hated the necessity for taking this wages job. He said he would work for himself if he could see a potential result but he hated working for a boss for wages.[33] At this time he had no choice. He needed money and the Annan River Tin Mining Company was his only option. That is, until he got a cheque from the *Bulletin*.

Around October 1912 Jack worked for only about a month at the Annan River Tin Mining Company. Right from the start of his employment Jack did not work too hard. The cry would go up, "Look out! Here comes the boss".[34] The three most notable facts that came out of that job were firstly that Jack's nickname "Cyclone Jack"[35] was confirmed when his workmates heard of it and ironically applied it to Jack's laziness. Secondly, it was while he was at Annan River that it became known that he was "Gouger" of the *Bulletin*.[36] Jack was actually dismissed because he was reading his published literary efforts when he should have been working.[37]

Thirdly, it was at the Annan River that Jack met his best mate Dick Welsh. At that time Jack was twenty-four years of age and Dick was only twenty but right from that day they became the best of mates. Jack had been learning a lot about the Queensland bush, first around Cairns then around Cooktown but Dick was a born and bred bushman. He had grown up with Aboriginal children and he could speak a number of dialects. When introduced by Dick,

Jack was able to observe many ceremonies and learn about customs that would otherwise have been kept secret.[38]

After (unsurprisingly) getting the sack he moved only about a kilometre or two downriver where he set up his own claim by borrowing equipment (without permission) from the company that had just given him the sack. He was successfully panning for alluvial tin at Rossville for about three weeks. At the end of his first week he had a full bag of black stream tin worth about £7 – more than double his wages at the Annan River Tin Mining Company. Then a downpour of rain flushed the creek and Jack's claim flushed with it.[39]

Jack moved on to Mt Finnegan with a couple of mates who were using a small hydraulic sluice to wash dirt off the mountain-side before shovelling it into a sluice-box to separate the tin. He was there for about three weeks in December 1912 and January 1913[40] but he soon tired of the "continuous bullocking toil" of Mt Finnegan and he went back to Rossville. There he condescended to accept another wages job this time with the Home Rule Company. Jack thought he could learn from the company's operations about hydraulic sluicing but after about a month (early in 1913[41]) he decided he had not learned much about tin mining but he had learned a lot about shifting heavy tree-stumps and logs.[42]

Some time in February 1913 Jack was working for himself again. With a mate he was blind stabbing for tin at Slaty Creek.[43] Blind stabbing was cold, wet, back-breaking work. Blind-stabbing sounds, to a non-miner, like the worst possible way to try to make a living. Chest-deep in cold water the miner works a long-handled shovel along the bedrock at the bottom of a stream. Bending low over the surface of the stream the miner feels for the small stones and gravel until he judges the shovel blade is full then he carefully lifts the blade clear of the water. Often the water washes the gravel off the shovel blade or the blade hits something on the way up and all the gravel is lost. This hard exasperating work did not suit Jack at all. He was working with a mate and although they were making good money, his mate shook his head hopelessly at Jack sitting smoking on the bank while his mate toiled in the freezing cold water.[44] Then a storm filled their claim with mud so Jack moved on.

Around March 1913 Jack moved to Mt Hartley about sixteen kilometres away where he luckily found an abandoned hut. In relative comfort Jack was

panning for tin in a small creek when he met the Baird brothers. The part-Aboriginal Baird brothers, Norman (twenty-five) and Charlie (twenty-three), had close family relationships with the tribes of the Bloomfield River and a wide knowledge of the Eastern Kuku Yalanji people and several other surrounding clans.[45] So for the next fourteen months (up to and including April 1914) Jack was not just prospecting. He was taking advantage of the Baird brothers' knowledge of Aboriginal people and their bushcraft to "learn the country". Jack finished *The Tin Scratchers* by acknowledging the vast amount that he did not know about the bush.[46]

In 1914, when Jack was nearly twenty-five, he left Dick in Cooktown and went prospecting for gold at Cape Melville with some other prospectors. He was at Cape Melville for about a month in August 1914 when Captain Dan Moynahan relayed the news that war had been declared. Jack travelled from Cape Melville to Townsville via Cooktown and Cairns to enlist but it took him nearly five frustrating weeks to get to the enlistment office in Townsville.[47]

When Jack heard the news about the war he was impatient to enlist. He could not wait for Moynahan's return trip from Point Stewart so he decided to walk to Cooktown. He waded mosquito and crocodile infested creeks for days before giving up and returning to wait for the *Spray*. He got to Cooktown to find there was no recruiting office so (with no money) he stowed away on the mail steamer *Musgrave*[48] but it was only going to Cairns and then returning to Cooktown. So there he was stuck in Cairns. Jack had no money and he had to depend on the Salvation Army food line for nearly a week before he stowed away again. This time he was discovered but he was hidden and fed by the crew.[49]

When Jack got to Townsville some time in the middle of October he found that the enlistment quota was full. He was disgusted that he had to compete with bigger, stronger men. Eventually, though, the skinny, sinewy bloke with the scraggy, dingy red beard was finally accepted for service in what became known as the Great War on 26 October 1914.[50]

Nearly six years later on Howick Island at the beginning of November, George and Jack had spent two weeks of companionship and being blown around by a gale. Then Jack said their relationship deteriorated; he said George had grown moody again.[51] As nature's storm blew itself out toward the end of November another storm was brewing. This time the storm was between George and Jack.

23

Second Fight December 1920

About a week or so before the end of November the weather was fine and sunny and the wind had moderated when a major crisis between George and Jack developed over lice in their blankets. One cooler night they each got an extra blanket. Both were restless during the night. George was awake and restless. When Jack asked him the reason for his sleeplessness he jumped out of bed and took his blanket outside.[1]

By the light of the wind-blown fire he examined his blanket and found lice. Angrily, shouting and cursing, he raced back into the tent and grabbed all his bedding and clothes. He threw the lot into the surf and waist deep, washed himself too. Jack stood outside the tent laughing and chuckling before collecting his blanket and settling down for the night under a tree. George ran back to the tent, dismantled it and washed it in seawater. He shifted all their belongings then lit a bonfire where the tent had been.[2]

According to his books, Jack seemed to take no part in this process and his amusement turned to anger as he watched George's frenzied efforts to cleanse the camp of lice. He watched as George taking down the tent, suddenly realised his clothes were being spread along the beach by the waves. Jack watched the naked figure of George chasing his clothes along the beach. He wrote that he vindictively regretted the sand-flies had gone.[3]

Next morning the wind moderated further and George was able to dry his gear. The tent dried too but again there was ill-feeling between them – a sort of tension that would not go away. For the next week if he was not fishing or sleeping, Jack spent most of his time up on the Peak watching for Moynahan. That suited George. He was not happy with Jack. He kept thinking about those dirty lice.

Finally, after a week of this brooding, George could no longer keep his feelings to himself. George knew he did not have lice in his gear before he came to the island and he was also certain there were no lice on the island before they came. So he was presuming Jack brought the lice in his gear.

November had gone and the two men had been three months on the island. Tension had been building steadily but the issue of the lice had really angered George. Early one December morning, when they were both sitting moodily by the fire, George decided to challenge Jack about the lice.[4]

The real crisis began when George asked Jack where he bought his blankets. Jack told George where he got his gear then George angrily accused Jack of bringing the lice on to the island. George's attitude was that Jack had polluted a formerly clean island with the parasites. George was angry. Jack said he had a "threatening" expression on his face. However, the dispute suddenly escalated. In the first book Jack angrily disputed George's allegation. In the second *Madman's Island* Jack scoffed at George; he said, in effect, that he could not care less if he had brought the lice on to the island.[5]

George's bottled-up anger turned to rage. All his feelings came out at once. He yelled at Jack and struck him in the face. Jack's retaliation turned the fight into an all-in brawl. In the first book Jack wrote that George's first blow knocked him down and George followed up with his hands around Jack's throat.[6] In the second *Madman's Island*, Jack after being struck in the face, jumped up and "*they went at it hammer and tongs*". Then the two men wrestled and fell to the ground striking and kicking and wrenching at each other.[7]

There is only Jack's word that George struck him first but even accepting that point, Jack could see how angry George was about the lice. He became angry or scoffed at the idea of lice in his blankets ("*...what odds?*"), so it could be argued that Jack should bear at least some of the responsibility for the fight. Even so it does not excuse George for what happened at the conclusion of the physical confrontation.

The fight was always going to be an unequal contest. Jack was sixteen years younger and in much better physical condition than George. As they rolled on the ground Jack repeatedly struck George on the temple[8] then fiercely punched George in the ribs.[9] Obviously George could not continue a fist fight. Indeed, had he known Jack had had some boxing experience, he probably would never have started the fight.[10]

At that point George did something that Jack (in both his books) consistently used to tag him with the "madman" label. Both men knew the rifle was in the tent but George was the one who dived for it. With a bit of quick thinking Jack realised why he ran for the tent. Jack raced toward the shelter of the mangroves. George fired at him twice.

It is now an appropriate time to consider the central tenet of Jack's two books. Both *Madman's Island* books are based on the assertion that after the two men were stranded on Howick Island, George went mad and tried to kill Jack. At this point in the story George had fired a rifle at Jack. Did George really want to hit Jack? The answer to the question is not clear but there's the fact – he did it. Jack said one bullet just missed hitting his head as it whistled past his cheek and the other hit a nearby mangrove branch.[11]

This certainly could be regarded as an indicator of madness and Jack said so in the first book: as he ran: "*his only wish was to get as far as possible from the madman's rifle*" and in the second *Madman's Island*: "*His face warned me that momentarily he had gone quite mad*".[12] The word "momentarily" is significant. Whereas in the first book George is said to be unequivocally mad and conflict between the two men was constant, in the second *Madman's Island*, George's periods of madness fluctuated.

Nevertheless, in both the *Madman's Island* books Jack deliberately set up the hypothesis that George was a "madman". This began as soon as Jack got back to Cooktown in February 1921. In his interview with the Cooktown police after he was rescued Jack was reported as saying George was "*out of his mind*", that he was "*strange in his head*" and that he was possibly suicidal.[13] Particularly in the second *Madman's Island* (up to the point that George fired the rifle) Jack used emotive words to re-inforce his hypothesis. Even at his second page he had George sombre and "growling" responses to Jack's happy optimism. On the same page Jack wrote that he smiled at his mate's grim face as George frowned toward the island. Early in the second *Madman's Island* Jack wrote that had been given friendly warnings about George's emotional instability.[14]

The first mention of madness in the second *Madman's Island* was when George found he left his medical supplies in Cooktown and insisted on keeping working for the rest of the afternoon. He said to Jack, "*I forgot the 'scope and I'll put up with it.*" Jack replied, "*You're mad*" to which George grimly replied, "*Not yet*".[15] With this exchange Jack was transparently setting up the scenario that George would go mad from time-to-time when the gas built up in his abdomen. Later, when George dashed a pannikin of water from Jack's hand and told him to get his own water, Jack attributed his behaviour to a build-up of gas.[16] At that point Jack said George was "*half-crazy*" and pondered the possibility of George descending into madness. Jack compared George's appearance to a crazed patient who, during the war, escaped from detention and leapt overboard from a hospital ship. Jack wrote that he thought about taking precautions and referred to George for the first time as a "*madman*".[17]

Superficially, the first fight occurred over a pannikin of water and the second fight over a few parasites in a blanket. But of course, even on Jack's say-so, the relationship between George and Jack was more complex and the motivation for George's violent behaviour cannot be easily categorized as "madness".

Jack said George's behaviour was the direct result of madness brought on by pain from accumulating gas in his abdomen. However, it is equally likely (again from Jack's version of events) that George's violent outbursts were culminations of steadily increasing frustration to which Jack's own personality and behaviour might have contributed. Whatever the true reason, a period of open conflict had begun.

With bullets from George's rifle cracking around him, Jack raced off into the shelter of the mangroves and on across the reef to the Mound. George did not try to follow him. So here's the situation in which they then found themselves.

For the rest of his stay on the island Jack had to live in a cave on the

other side of the kilometre-long reef that was safely passable only at low tide. George was living in the tent at the foot of the Peak and the well was nearby. And they were at war. It was survival of the fittest and (from the fight) George would have known absolutely that he was not the fittest. He was no match for Jack's youth and strength.

24

First Raid and Retaliation December 1920

George would have known Jack would be planning his next move. Just as Jack wrote in both of his books; his main problem was lack of water but neither did he have fire. They had long ago run out of matches and although it is possible (but difficult) to start a fire without matches, keeping a fire going was essential. George had the fire and the water. He would have known Jack would be planning an attack.

Today, over ninety years later, it is easy to see how the impasse could have been resolved. Perhaps George could have apologised. Perhaps George could have walked across the reef, rifle at the ready and yelled to Jack that he could get the fire and water he needed. Perhaps. Perhaps. Back then, stranded on Howick Island, George decided to defend himself.

Each of them, both ex-servicemen, knew the other was planning something. Jack had to mount a raid to get water and fire. George would have been trying to think when and how he would come. George had the rifle. Jack had agility and strength.

Jack was thinking through his options. He knew he could make fire but the only source of water was on George's side of the island. He also realised that his fish-spear was still in the tent. Jack spent a miserable cold night and early next morning he decided he must take action.[1] Jack half-expected George to have recovered from his "temporary madness" but he was wary. George could have been waiting in ambush by the well.

However, it was not only Jack who would have been doing some planning. To eat George had to catch fish and he could only fish at low tide, just when his adversary might come creeping through the mangroves. Even so, he could spear crabs at high tide so he would not starve. George had another problem that Jack could exploit. Soon he would be extremely vulnerable when he had to irrigate his bowel.

Like Jack, George could not afford to guess wrongly about the intentions of his adversary. George had suffered a severe beating from Jack and he would not have wanted this to be repeated. He would also have been conscious of the fact that he had fired a rifle at Jack. He could not expect Jack to easily forget this hostility.

So George had to guard his territory. He could not be at the campfire and at the same time monitor the well. He might have tried patrolling but he soon realised the impossibility of maintaining this level of vigilance. George decided to guard just the well. He needed to know if Jack was about and he needed to know his intentions. George kept hidden and watched for Jack's arrival.

For his part Jack had decided to creep through the mangroves when there was still half a metre of water over the reef. Chancing sharks, he had to half-swim and half-wade across creeks racing with the outgoing tide. He came out of the gloom of the mangrove forest into sunlight at the foot of the Peak. The well was just inside the edge of the mangrove forest about 275 metres to his left. Jack ran along the edge of the mangroves until he was about ninety metres from the well. Then he ducked into the mangroves and crept stealthily forward.[2]

Jack was already hiding in the mangroves watching George as he walked to the well to get a drink. After he had a drink George checked for tracks to see whether Jack had been there. There were no tracks so he climbed a tree to guard the well. This was a mistake. Jack was watching.

When he saw George climb the tree Jack hurried through the mangroves and around the Peak to the campsite. Inside the tent Jack found the kerosene tin full of water as well as three full billycans. After he had a drink he grabbed his fish-spear and made a swag of personal belongings wrapped inside a blanket. He passed the spear-haft through the handles of the water containers. Then with his swag and a fire stick he started back to the Mound.

In the second *Madman's Island*, Jack wrote that it was on this raid that he found Tarquay's Log of the *Sea Foam* but this might not be correct.[3]

Some time later George would have tired of sitting in the tree. When he got back to the tent he would have found Jack had raided it. In itself this raid was not a big problem. Jack had only taken his own fish spear and his own bedding and personal possessions. It really should not have worried George that he took the reserves of water or the billycans – Jack had bought them in Cooktown[4] but George would have been upset that Jack had so easily beaten his security. Angrily, George set out across the reef to have it out with him.

Jack had hidden the water cans in a crevice of rock and started a fire in his cave. He had time to catch and cook a couple of fish. He was just finishing his meal when he saw George coming with the rifle. Jack saw George approaching rapidly and waving his arms. From this body language he inferred that, "(George) *wanted those billycans and everything else* (he) *had taken*".[5]

In the first book Jack took a really aggressive stance as soon as he saw "the madman". In that book Jack wrote that, "*a sudden furious anger took hold of him*" and it was not George who was waving his arms. It was Jack who was shaking his fists at George. Jack's attitude was: "*He's after my life; I'll be after his*" and he deliberately lured George into a trap.[6]

Now Jack had reason to be apprehensive. George had tried to shoot him and Jack had raided George's camp, even if he had only collected his own belongings. He could see George approaching across the reef holding the rifle. However, the only indications of George's malicious intent were that he was approaching rapidly and that he was waving his "*threatening arms*".[7]

George might not have intended to kill Jack. It is quite feasible to assume he wanted to talk but that he had the rifle as a deterrent against Jack's superior strength and fighting ability. George might have wanted to tell Jack what he thought of him and the situation in which they found themselves. To be fair, however, for Jack to have made this assumption might have been fatal.

Jack did need to take precautions but his subsequent actions went far beyond merely defending himself.

There is a reasonable inference that George did not have homicidal intent. He came openly across the reef. As a trained soldier he would have known the need for a surprise attack. If George had been trying to kill Jack he would have used the rifle to its best advantage – from a distance. Getting really close to the enemy before using the rifle is not the best strategy. If George had been deadly serious he would have crept unseen to within range. Then like a sniper, he would have shot Jack.

Jack recognised this fact. In the first book he wrote that George took no cover because his "*twisted brain*" told him the rifle made him "*invincible*".[8] Of course George would not have thought himself invincible and as a trained soldier experienced in sniping, Jack knew it. As it turned out, George's true intentions did not matter. Jack tried to kill George.

By showing himself on top of the Mound, Jack had lured George to a place where he had prepared for an attack. Hidden, he waited in ambush until George started to climb. George's first indication of Jack's deadly intent was when a large rock shattered into fragments directly beside his head. Jack wrote that the rock hit about a metre above George's head – close enough for his face to be cut by flying splinters of rock. Jack only just missed killing George with that first rock and then in rapid succession he showered stones down on George as he was trying to scramble over the boulders at the foot of the Mound.

There was no doubt about Jack's intent. Jack let George, "*get about* (7m) *below the boulder on which he was lying. Then, leaning over, he took as careful a balancing aim as possible with the largest stone*".[9] Jack was dancing around and laughing at George as he hurried back across the reef. When he got there – a kilometre away – George fired the rifle in a futile gesture of angry frustration.[10]

The next two weeks of December passed by uneventfully. George did not try to talk to Jack. Even though in the first book Jack said the water he got from his first raid lasted for two weeks, he did visit George's camp on several occasions.[11] Jack went to the well to get water without obstruction from

George. Jack thought George was watching from the mangroves but he could also have been busy tending his vegetable garden or improving the campsite. George gathered stones to build up and level the area around the tent and he added more stones to the wall of the well. He would not have been willing to risk another King tide polluting the only source of fresh water on the island.

The fishing was good for both men and the weather was perfect. More importantly, George's system of irrigating his bowel must have been working reasonably. At least he did not die from it.

At the end of the first two weeks of December, the weather changed. Neither George nor Jack was surprised by the black storm-clouds that come with the Wet Season but the clouds came with a thick fog. The strange aspect to this weather was that the stormy clouds and fog would obscure everything for about twenty minutes then clear to bright sunshine before closing in again. Then Moynahan's *Spray* turned up at last.

25

Moynahan Sailed Away December 1920

For months Jack had been anxiously anticipating Moynahan's arrival. Right from their first few days on the island Jack had been saying he was anxious to get back to Cooktown.[1] At the end of their first month when George and Jack were rationing their tobacco and Jack was disappointed they had not found much tin he was even more anxious to leave.[2] Even when he and George were getting along quite well, they were never the best of mates and Jack was anxious to put this episode of his life behind him. George's thinking was quite different. He wanted to stay on the island.[3]

Days turned into weeks and then into months and still the *Spray* did not appear. This is one of the main questions unanswered in either of Jack's books. When Moynahan did come back he was months overdue. Why was he so late and why, when he eventually did come back, did Moynahan not take the two men off the island? In the first book Jack does not discuss these questions but in the second *Madman's Island*, Jack blamed George for Moynahan's non-appearance.[4]

Jack said he learned long after his rescue from Howick that Moynahan did not pick them up because George had deliberately instructed Moynahan not to bother unless there was a well-defined signal. The inference is that George and Moynahan knew this signal but knowledge of it was withheld from Jack. In neither book does Jack explain this peculiar allegation. It does not make sense that George would have made a secret arrangement with Moynahan and it does not make sense that George would have withheld the signal from Jack when the *Spray* eventually did arrive.

Another strange factor is that just before the *Spray* appeared both men had tried to kill the other. Surely (if there was a special signal) George would have given Jack the signal to get him off the island. The allegation of a secret pre-arranged signal does not explain Moynahan's lateness and it is not a reasonable explanation for his failure to see Jack's signals. The simplest explanation is most probable.

For years Moynahan had traded up and down the coast but mainly between Point Stewart and Cooktown. He would do this trip (depending on trade) about every three months.[5] To minimise the risk of cyclones Moynahan would have tried to avoid the height of the Wet Season. Moynahan delivered George and Jack to Howick early in September and he agreed to return in about a month. He was over two months late. When he finally got back to Howick in the middle of December he was sailing back to Cooktown.

The reason for Moynahan's lateness might have been the weather. He would have encountered exactly the weather George and Jack endured. For all of October the weather was dead calm. Moynahan could not have sailed the *Spray* but he could have used his motor. He might have decided not to use his fuel and as soon as the calm passed there were gale-force winds for much of November. The simplest explanation for Moynahan's delay is that he might have waited out the calm and the gale before heading back to Cooktown.

When Moynahan did get back to Howick he did not see Jack's frantic signals but there was a thick swirling fog that would have reduced Jack's visibility. Further, Moynahan might not have been looking for a signal. Because he was so late he might have assumed that the two men had already gone. In any case, because it was late in the year he might not have been too keen on waiting around Howick Island. Of course this explanation is pure conjecture but it makes more sense than Jack's explanation of a secret signal.

It was about midday when Jack first saw the sails of the *Spray* across toward the mainland at Cape Melville.[6] All that afternoon Moynahan was tacking directly toward Howick. He arrived off Howick late in the afternoon and he sheltered behind what was probably Newton Island. Jack had to wait all of that afternoon until the tide dropped enough for him to hurry across to George's camp to tell him about it.[8] In the first book he did not go to tell George.[9]

In the second *Madman's Island* Jack wrote that he was terrifically excited and ran across the reef to George's camp. He spoke to George as though nothing had happened between them. The chance of escape had driven all the unpleasantness of the past eleven weeks from his mind. George was not so forgiving. Only a couple of weeks earlier George had fired his rifle at Jack and (later) Jack had tried to kill George. The rock-splinter scars on George's face might have healed but the incident obviously still stuck in his mind.[10]

Jack excitedly told George about the imminent arrival of the *Spray* but George said he would not be going. Jack did not care; all he could think of was escaping from that "*wretched island*".

Next morning George still had made no preparation to leave. He firmly declared his intention to stay.[11] In grey, overcast weather Moynahan sailed through the channel between Houghton and Howick Islands. This was hazardous because a swirling mist frequently reduced visibility. Jack, with George, had been watching from on top of the Peak but when he saw the *Spray* would sail close by the Mound he raced across the reef waving his tattered shirt.

To avoid the dangerous reef Moynahan sheared off and anchored about 1.5 kilometres away near Coquet Island. The fog kept closing in and hiding the ketch. Jack was frantically waving his shirt from the Mound but still the fog defeated him.

Jack was bitterly disappointed when the *Spray* took advantage of a slight breeze and cleared the reefs and sailed off toward Cooktown. Jack said he nearly went crazy as Moynahan sailed away. In the first book Jack wrote that he rushed into the surf and was thrown against shell-encrusted boulders then, dragging himself away from the water, he lay like a sick animal all through that night.[12]

Christmas! But there would have been no celebration on Howick Island. George would not have cared for celebration and Jack would have been in a filthy mood when Moynahan sailed away. The cold, stormy weather continued to bring black clouds and fog with the constant possibility of rain. Jack had gone back to his cave on the Mound. But not for long.

26

Second raid and retaliation December 1920

Toward the end of December Jack mounted another raid on George's camp but this time he escalated hostilities. In neither book did Jack adequately explain why he carried out such a provocative raid. Jack set fire to George's tent.

It is unclear whether Jack destroyed all of George's belongings as well as the tent. In the first book Jack wrote that he crept through the mangroves to George's camp then taking a fire-stick from the campfire, "*he deliberately placed the blazing stick against the tent side*".[1] In this version Jack burned the tent and everything in it! In the second *Madman's Island*, Jack wrote that he threw George's belongings outside before burning the tent.[2]

In either version this provocation was quite unwarranted. It was the day after Moynahan's *Spray* had departed.[3] As the *Spray* was approaching Jack had visited George's camp not once but twice. Although George was surly there was no suggestion of animosity and Jack gave no indication that he was fearful.[4]

Yet Jack described creeping through the mangroves instead of walking openly into George's camp as he had done for the two days earlier.[5] Jack wrote that George was guarding the well[6] but this was an assumption. Jack could not see George but the birds around the well were uneasy so he assumed George was there. Why, given his visits of the last few days would he come to this conclusion? Whatever the reason, Jack did make this assumption so he cautiously crept away through the mangroves and around the Peak to

George's camp. He set fire to the tent (and, perhaps, everything in it). Hidden in the mangroves, Jack waited until he could see George, "*racing up the path from the mangroves, waving his rifle and screaming*" before he went to the well for water. Writing in the first book, Jack then burst into gleeful laughter.[8] There is no rational reason for his glee. Nothing had happened in the meantime except that the *Spray* had sailed back to Cooktown without Jack. George had done nothing to excuse Jack burning his tent.

In the second *Madman's Island* Jack justified his action by asserting George was forcing him to live like an animal in a cave. He reasoned that without the tent, George would have to find himself a similar den.[9] Whatever Jack's reasoning he could not expect George to ignore this blatant provocation. Just as Jack had been watching George from time-to-time, George would also have been keeping an eye on Jack.

Jack took his store of precious water back to the Mound and hid the containers in a rock crevice. Obviously, George had seen the hiding place that Jack had found for his water. Carefully at low tide, George crept through the mangroves at the edge of the reef. He knew Jack would be fishing the creeks. George crept around the Mound to Jack's water store and emptied the cans. Then he filled them with sand.[10] This was, of course, just some more vindictiveness but it was direct tit-for-tat retaliation for Jack's burning of George's tent.

George would have had no way of knowing at the time that, while he was destroying Jack's water store, Jack had been bitten by a shark. While wading in a mangrove creek Jack narrowly avoided a really serious injury. For Jack the loss of his store of water meant a much more significant effect than George had intended. Jack had just speared a cod when he heard a swishing noise behind him. Glancing back he saw the fin of a shark speeding toward him. He jumped for the mangrove branches but he was not quite quick enough. The shark's teeth had just grazed his leg but still made a nasty gash. Jack tore off his shirt and tightened a rough bandage around his leg until the rotten cloth tore. He then made a pad of his shirt and held it against the gash.[11] Back at the Mound he discovered his water cans filled with sand. Jack's leg was stiff and throbbing. He felt sick and faint. With the loss of his water he was badly scared. Before George had to run back along the reef to beat the tide, he was

laughing at Jack and relishing his revenge.[12] He did not know about Jack's injury.

Jack's luck took a turn for the better when that night there was a heavy shower of rain. He emptied the sand from his water containers and soon they were filled with fresh rain-water. Every rock crevice was also filled with water but Jack had another problem. Just as the shark swished past in the mangrove creek he had speared a fish and (in escaping the shark) he had lost his spear.

Painstakingly he set about making another spear with wooden prongs instead of the iron tips of the spear he had lost. He hardened the wooden tips carefully in the flame of his campfire.[13] The wooden spear tips were not really satisfactory but for the time being Jack could not chase fish. He was hobbling about using a stick for a crutch. Until the gash in his leg healed he decided to stay out of the water. So Jack speared some crabs and collected shellfish to eat. The shellfish meal was a mistake. To add to his misery he was then violently ill.[14]

The old year, 1920, ended with weather that was wet but uncharacteristically cold and miserable. Jack's birthday had come and gone in September and Christmas had passed unremarked. Jack was recovering from the nasty gash on his leg. Although George and Jack were both fishing successfully it was now possible to fish out some of the creeks. At the end of the year the two men had been on Howick Island for four months and their relationship was dreadful.

27

Third Raid
January 1921

The New Year, January 1921, inexplicably brought with it a change in the relationship between George and Jack. With one dramatic exception (when George had to struggle free from a mud-pool), January was notable for the fact that the two men talked freely and often. Even so, the New Year probably passed unnoticed by either of them. Jack gave no indication in either of his books that he was keeping track of the date. Indeed he wrote. "A*ll count of days and weeks was lost.*"[1] However, the New Year was marked by Jack making a surprise visit to George's camp.

Jack did not mention this visit in his first book. Instead he writes as though the men were still constantly at each other's throats.[2] In the first book Jack seemed to use the guerrilla war assertion to support his claim that George was mad. In the second *Madman's Island* Jack wrote that (except for December) the two men talked frequently and Jack's notion of George's madness was linked to gas building up in his abdomen.

Of course, George would have known that Jack was occasionally coming across the reef and climbing the Peak. Jack had been to collect water and later he had re-built a signal cairn on top of the Peak.[3] Jack had also seen the hut George had built to replace the tent Jack had burned. George had built a little hut from driftwood and thatched it with grass. As Jack said, losing the tent was not a disaster for George. Even on their first day on the island, George suggested building a hut out of driftwood and Jack had replied, "*We won't want it; we'll be gone in a month.*"[4]

Jack was also aware (even though he denied it in the first book[5]) that he had watched George tending his vegetable garden. George already had sweet potatoes growing and he had seeds for pumpkins and melons too. So, Jack strolled across the reef, as calm as you like with no sign he was scared of the "madman" and the madman's rifle. If he was nervous he did not show any evidence of it. George and Jack talked freely about fishing and about Jack's wooden spears. George offered him some wire to make iron prongs for his spears. George also offered him some of the sweet potatoes.[6]

Over the next ten days Jack walked across to talk with George quite often.[7] In his first book Jack does not mention this pleasant period; probably because it does not fit his "madman" theory. In the second *Madman's Island*, although Jack said there was plenty of time to yarn with George, he did not go into detail about the subjects they discussed. Even so, during that time Jack's experiences of war in Palestine did come into his mind.[8]

It is inconceivable that the two ex-soldiers did not at some point yarn about the Great War. This ten days, the first ten days of 1921, would have been as likely as any other that Jack would have told George about his war. George would have been more circumspect. He seems not to have explicitly told Jack that his operation was not the result of a war wound but he seems to have allowed Jack to believe it was so. Even after he got off the island Jack still believed George was suffering from a wound in the stomach which he received in the war.[9] This was not true. His operation was not a war wound but after the war, back in Australia, his pre-existing bowel condition was ruled as being exacerbated by his war service.[10]

28

War

During the First World War more than one in three of the total male population between the ages of eighteen and forty-four enlisted. At forty-four years of age George joined up in Cairns on 26 April 1917.[1] It is not known why he did not enlist earlier. The age requirements in August 1914 were eighteen to thirty-five years of age but the maximum age was raised to forty-five in June 1915. It is possible that George might have been initially rejected for some health reason. Early in the war about a third of all volunteers were rejected but as the war progressed standards were relaxed.[2]

George's unit was posted to the 31st Battalion, 14th Reinforcement and his unit sailed from Sydney on board HMAT A20 *Hororata* on 14 June 1917. On 27 December 1917 he sailed from Southampton to France to reinforce the 41st Battalion AIF. George might have seen service with the 41st in Belgium early in 1918 but by March of that year the Battalion was active in France when the German army began a major make-or-break offensive. George would have fought with the 41st against at least one major German attack in March 1918. Predicting the arrival of American troops the Germans were trying to gain a strategic advantage. Although the German troops were exhausted from recent fighting on the Eastern Front this was a period of intense conflict. There were many casualties on both sides.[3] George was not one of them.

As the 41st battled against the German offensive, on 14 April 1918 George was admitted to hospital. Four days later he was discharged back to the 41st but he was still sick. After another period of hospitalisation in May he must have been assessed as unfit for active service. On 9 June 1918 he was transferred from the 41st Battalion to the 4th Division Training. George was on leave in England from 9 November 1918 to 23 November 1918 when he was admitted to the Military Hospital at Devonport "seriously ill" on 17 November 1918. During December 1918 George was still seriously ill with an "intestinal obstruction".[4]

Early in 1919 George had a laparotomy operation for chronic obstruction of the intestine. Later it was said George had:

> "*Chronic constipation and trouble with bowels on and off for some years – became pretty bad last June and evacuated and returned – came over on leave and reported. Has had ileectomy, colostomy and appendicostomy done. Washes his own bowels every other day. Is getting a belt to* (stop?) *faeces coming through wound now.*"[5]

An ileectomy is the surgical removal of the final section of the small intestine. In a colostomy procedure the end of the large intestine is drawn through the abdominal wall and stitched into place. This opening (usually in conjunction with a stoma appliance) allows an alternative for faeces to leave the body. An appendicostomy permits irrigation by water to improve colonic evacuation – to irrigate and drain the large bowel.

Understandably, after this operation George, felt low-spirited and he was easily tired. His doctors decided no further treatment would be provided so, after a period of convalescence (including another hospital admission) in England, George returned to Australia "Invalid" on the *Marathon* on 19 April 1919.[6]

Jack's war was quite different. Jack enlisted on the 26 October 1914. He said he was one month over his twenty-seventh birthday but in fact he was one month over his twenty-fifth birthday. He became Trooper 358 in the 5th Light Horse AIF. He served at Gallipoli for about two months where he was twice hospitalised. He served in the Middle East for about twenty-one months before being wounded for the third time. He was discharged on 2 January 1918.[7]

Neither Jack nor George distinguished themselves by promotion – both remained private soldiers throughout their service. Neither won medals for

valour. There is no way of knowing whether they were otherwise effective fighting men but Jack did write about the concept of fear. Despite Beverley Eley's opinion that Jack "*was not fearful for his own safety*"[8], Jack wrote frankly of his fear and discussed the characteristics of fear in battle.[9] But as Eley elsewhere wrote, his personal courage was undoubted. Eley quoted Major Jack Cain who recalled how Jack suggested that he could destroy a dangerous Turkish gun nicknamed by the Australians "Beachy Bill". Eley wrote, "*...the Turks were using the gun with great efficiency and causing many fatalities and casualties among the Anzacs. Jack was commended for volunteering though he was refused permission to undertake the dangerous mission by Colonel Harris who felt the position was too well guarded*".[10] Eley also quoted Frank Byron of the 5th Light Horse who said that Major Cain often spoke of Jack's suggestion and wondered if it could have been successful. Major Cain said Jack would have to have been gamer than him to try to take out the gun and this from a Major who won the Military Cross![11]

Later, Jack wrote about war and strategies for defending Australia but except in *The Desert Column* he rarely mentioned his personal experiences of war and then mostly indirectly. For example, only once in *The Desert Column* did he directly describe killing a man. In most other passages he writes as just one man in his section, describing the killing almost neutrally as in: "*It was grand shooting, those waves of men who came on and on and on. As their ranks thinned other men sprang up from the bushes and refilled the lines.*"[12] The one direct mention of killing in the book is his account of seeing a Bedouin appear over a sand ridge only about six metres from where he was lying. Jack aimed carefully and shot him dead.[13]

In *The Yellow Joss*, a collection of short stories published first in 1934, he also describes killing a man – this time a Turkish sniper. After carefully stalking his man through a field of barley, Jack took aim "*at the little hollow below the throat*" and fired. Jack wrote that he bounded out of the barley and was on the spot even before the Turk had rolled over, dying.[14] Jack claimed this story was factual but there is no mention of such an event in *The Desert Column*. In similar dramatic style in the Daily Telegraph Pictorial on 12 November 1929 Jack wrote about shooting at and then bayoneting a Turkish soldier.[15]

Jack experienced the horror of Gallipoli immediately upon his arrival on the transport ship Lutzow in May 1915. On only his second day (still on board the Lutzow off Cape Hellas) Jack watched as Turkish artillery devastated a battery of four Allied guns being drawn by galloping horses. A shell burst in the middle of the galloping battery. The first and last of the battery galloped on but the centre was gone. Then they watched the retaliation. A cruiser raced close inshore and wheeled around to pound the fortified artillery into silence. Jack described the sound of the cruiser's guns crashing and echoing off the hills. On just his third day Jack and his mates were just about to board their landing craft when Jack saw another landing craft full of soldiers blown up by a direct hit from an artillery shell. What a welcome! And they had not yet got off the Lutzow.[16]

Their landing was put off for another day but then they landed safely under heavy shell and small-arms fire. Strangely, Jack and his mates were glad to be on the beach. His first stop was the infamous Shrapnel Gully thundering from so much gunfire that talking was impossible. Jack described rifle bullets chirping through the bushes like busy canaries. Death was all around. Jack found a "bonza" pair of near-new Turkish boots but when he pulled at one – out came part of a Turk's leg.[17]

Still a novice soldier and still trying to absorb the dreadful reality of war, Jack was surprised at the sense of anticipation in the ranks of soldiers – a sense that anything could happen at any minute.[18] Then in an odd contrast, in the midst of overwhelming sound, massive confusion and imminent death, in the lull between bombardments Jack and his mates went swimming.[19]

Light-hearted interludes like this were few. It took Jack only a day or two to become sickened by what he saw (and smelt) of war.[20] War for Jack was now constant rain, mud everywhere and the stink of corpses rotting just in front of their trenches. Thousands of them.[21]

Although he could not then know, Jack still had years of battle ahead of him. Through it all Jack kept his diary. It was not easy. At one point, still at Gallipoli, he described writing while lying on his back with the muddy ground rocking from shell fire and earth tumbling from the damp walls of his dugout.[22]

Only twelve days after arriving at Shrapnel Gully Jack had a lucky escape

from sudden death. Not so lucky was the man immediately in front of him. Walking for water the man was shot by a sniper. He suddenly flung up his water bottles, wheeled around and with a startled look of surprise fell dead at Jack's feet.[23]

On 30 May 1915 Jack's first hospitalisation began as just a minor incident. A piece of shrapnel grazed his knee. Only a scratch – he thought! But not so. His leg became badly infected and he was taken off to the hospital ship *Franconia*.[24] It was nearly a month (28 June) by the time he arrived at the Egyptian Government Hospital at Alexandria. All through July and much of August 1915 Jack was off the Peninsula. In one sense this was a truly lucky wound.

Lucky, because on 7 August 1915 over 600 of Jack's mates from the Light Horse were slaughtered. Bayonets fixed to their rifles (but no bullets for the rifles) the first wave was ordered out of the trenches across thirty metres of open ground. This near-criminal decision meant those young Light Horsemen had to charge securely dug-in Turkish machine guns! The first wave was cut to pieces and so was the second. Even the Turks yelled for the carnage to stop but still the same fate befell the third wave. Jack was lucky that he was in hospital in Alexandria.

Of course, Jack's respite did not last. On 23 August Jack was on board the A25 *Hunstead* heading back to Gallipoli.[25] A few days after he got back (perhaps on the morning of 30 August) Jack spotted for the renowned sniper Billy Sing.[26] There was some irony in this because much later Jack wrote of his hatred of snipers: "*We hate snipers ... A sniper just kills as many men as he can from ambush, then when surrounded simply holds up his hands.*"[27]

Then by September 1915 Jack was sent to Lone Pine – the most dangerous place in the whole Gallipoli line. Jack wrote about the stench of half-buried bodies in the floors and walls of the trenches. He said these corpses made the sandbags greasy. Men were killed there every hour. Jack's words in *The Desert Column* are still frequently quoted to illustrate the horror of the place:

> "*Maggots are falling into the trench now. They are not the yellow squashy ones; they are big brown hairy ones. They tumble out of the sun-dried cracks in the possy walls. The sun warms them I suppose. It is*

beastly. We have just had dinner. My new mate was sick and couldn't eat. I tried to, and would have but for the flies. I had biscuits and a tin of jam. But immediately I opened the tin the flies rushed the jam. They buzzed like swarming bees. They swarmed that jam, all fighting amongst themselves. I wrapped my overcoat over the tin and gouged out the flies, then spread the biscuit, held my hand over it, and drew the biscuit out of the coat. But a lot of flies flew into my mouth and beat about inside. Finally I threw the tin over the parapet. I nearly howled with rage. I feel so sulky I could chew everything to pieces. Of all the bastards of places this is the greatest bastard in the world. And a dead man's boot in my firing-possy has been dripping grease on my overcoat and the coat will stink forever."[28]

As recently as 2006, in an address by His Excellency Major General Michael Jeffery, AC CVO MC, Australia's then Governor-General, used this passage from *The Desert Column* at a memorial service at Lone Pine, Turkey.[29]

To give him a rest, Jack was sent to a trench about fifty metres back from the front line and things got even worse. It was dreadfully overcrowded – forty-eight men in a trench just half a metre wide. It was dark from overhead iron rafters supporting sandbags. The air was stagnant and stinking from the bodies of dead men thrown up on the parapet with the sandbags. Jack wrote, "*What ho, for the Glories of War*".[30]

Jack passed his twenty-sixth birthday at Lone Pine. In his diary Jack did not mention his birthday. Around his birthday (perhaps even on his birthday as a macabre birthday present) Jack was wounded a second time. He was wounded in the arm when a Turkish bomb landed in a trench near him. Jack threw his overcoat over the bomb and tried to back away but there were too many men pressing behind him. Directly in front of him, his mate caught the blast and later died of the wounds he received.

Jack said he was not badly wounded but there he was back on a hospital ship – the *Salta* – bound again for the Egyptian Government Hospital at Alexandria and then a period of recuperation at Ras-el-tin convalescent home. This marked the end of his time at Gallipoli.[31] No doubt Jack was pleased to be away from that horrible place. While he was recuperating Jack missed the

retreat from Gallipoli. The last sentence of his diary entry for 3 January was, "*We could not believe the news of the Evacuation*".[32]

By February 1916, Jack was on duty again. However, away from the mud, flies and maggots of Gallipoli now his world was sand, more sand, flying sand and sand everywhere.[33] Jack was so bored that he was envious of men heading off to France.[34] Although Jack could hear guns and see searchlights (he was told the Turks were only a day's march away) it was well into April before he came anywhere close to some action.

A force of Turks and Arabs had attacked a British camp. Jack's mob was sent to re-inforce them. Jack was disappointed that the Turks had fled before he and his mates arrived. For the next few months this pattern continued. Jack whinged about heat, sand, flies, sand, wind, sand and so on. Constant training in the desert had men and horses to a peak of fitness capable of enduring just about anything the desert could throw at them. Jack thought they were at least the equal of the Bedouin. He and his mates seem to have been constantly riding out on fruitless patrols after false reports from the despised aeroplanes. Or like on 1 June, when they rode all night to re-inforce the New Zealand Mounted Rifles. They were too late and rode back the next night. Jack succinctly summed up his life in the desert: "*This cursed desert monotony is nerve wracking: we are the lost legion...Curse this inaction.*"[35]

Finally he got his chance. In July 1916, Jack was chosen with a small party of men to form a little base camp about thirty-five kilometres in front of their own lines in order to spy on a big Turkish camp. There were no diary entries until it was all over. On 31 July Jack wrote that he wished he could more adequately describe what he called the most exciting trip of his life. Mainly, Jack's patrols were told to avoid contact but inevitably they were discovered. On this occasion the Turks had laid a huge ambush. Jack thought it looked absurdly out of proportion, "*almost a whole regiment had been turned out to catch one tiny patrol.*"[36] To get one final look at the troop numbers Jack, his lieutenant and another man climbed a razor-back sand hill to look directly down on to massed Turks. Then, only about six metres from where Jack was lying, a black clothed elbow appeared above the ridge of sand. Jack saw a rifle muzzle and then the head of a Bedouin. Three times he shot that man then there was a frantic gallop to escape.[37]

There was near-disaster when Jack's horse fell and rolled on him. Jack's Sergeant came back for him. Jack's head was buzzing as he tried to steady his frightened horse. The Sergeant calmed Jack and talked him back on to his horse. The first bullets were hissing past him as he raced after his mates with only one foot in a stirrup. The other had doubled up and tangled with a feed-bag when the horse fell. It was a race for life down the hill as their horses galloped and rifle bullets zipped past. Jack said he could not hear the rifles only the zip of the bullets. He said he and his mates were laughing with excitement as they rode away.[38] Back at their base camp they were relieved by New Zealanders who were shortly after overwhelmed by the Turks and taken prisoner. Jack was glad it wasn't him in a prisoner-of-war camp.

Jack also missed the furious fighting around Romani and El Katia but he describes magnificent acts of heroism and dreadful losses on both sides from massed bayonet charges and point-blank night-time gunfights. He expressed admiration, widely-held amongst Australian troops, for the willingness, determination and bitter stubbornness of Turkish soldiers.[39]

Jack had been bored and itching for action in the first six or seven months of 1916 but in August he got all the action he could take. The action began on 5 August when Jack was part of the re-inforcements sent to El Katia. As they drew nearer, the sound of the guns grew louder and they could see shrapnel bursting over many kilometres of the El Katia oasis system. A bayonet charge was ordered on a battery of heavy Austrian guns thought to be hidden in an oasis. Bayonets fixed, they were off, first at a canter knee-to-knee, then in a wild gallop toward the oasis.[40] Leaving the horses under cover they headed off over open ground. Machine gunners opened up on them.

Jack said his legs felt as if they were made of lead:

> "*It is a terrible nightmare when a man strains his very heart out and cannot move an inch. The air was just one sizzling hiss – flying bullets at point-blank range possess an awful sound. At the bushes' edge I flung myself against a sand-clump in frantic fear.*"[41]

The Austrian guns were not there but Jack and his mates gave the Turks and Germans a hammering. Inexplicably (from his viewpoint) just as Jack expected a bayonet charge to finish the job, they were ordered to withdraw.

Toward the end of August 1916 Jack spent some time in hospital with dysentery and brought his diaries up-to-date. By September (another birthday in battle – his twenty-seventh) Jack's hunger for action had abated. Jack wished the war to end[42] but (although he could not know it) he still had all of the next year to survive before his war was over.

After a period of leave late in 1916, the New Year came in quietly. Jack was getting plenty of patrolling in enemy held territory. For instance, late in February he was helping support New Zealanders, guarding them from Turkish infantry entrenched at Shellal and skirmishing with other cavalry patrols from the Turkish side.[43]

Then late in March 1917 Jack saw thousands of infantrymen die needlessly at Ali Muntar near Gaza. On a British General's orders, rather than the Desert Column and the other mounted soldiers, infantry were to take Gaza. Ali Muntar was taken and lost three times then taken again at bayonet point.[44] Jack's section was higher and at the edge of the main action when they came under heavy rifle and machine gun fire from huge three-metre high prickly pear hedges. They galloped right up to the hedges and jumped off their horses – leaving them in charge of horse-handlers – then on through the prickly pear into hand-to-hand bayonet fighting.

Now here was Jack (the man who only a few months earlier had been complaining of boredom), springing at the closest Turk to thrust and spring aside, thrusting again and again. Jack was right there, face-to-face with (as he wrote), "*the grunting breaths, the gritting teeth and the staring eyes of the lunging Turk*" and getting more action than he needed. He wrote in his diary of the "*sobbing scream as a bayonet ripped home*".[45] Jack was getting his action! How he now thanked all those previously cursed hours of drilling and training:

> "*Amateur soldiers we are supposed to be but, by heavens, I saw the finest soldiers of Turkey go down that day, in bayonet fighting in which only shock troops of regular armies are supposed to have any chance.*"[46]

In 1917, although Jack's section seemed to be more often in reserve, on observation duty or on the flank of major battles, there were often times when he was fighting for his life. For some of these times it is necessary to read between the lines of *The Desert Column* but other incidents are explicit.[47] Around 20 April (understandably Jack seemed to have lost track of dates) his squadron of 100 men was surprised by a thousand Turks. After some confusion as to whether they would be required to fight or run, Jack's squadron was ordered to hold their positions to the last man. They tumbled back into shallow trenches and prepared to fight to the last.

Jack wrote that he knew it was all up, that they were hopelessly outnumbered. In the last rush every man would be bayoneted at his post.[48] He made up his mind that he was going to die and determined to do the hero act. But in a surprising anticlimax when Jack's small squadron opened fire a thousand Turks retreated. So Jack did not become a hero that day – dead or alive.

Then on 2 November Jack was seven kilometres from and within sight of Beersheba. In all of history there has been no greater charge of mounted men and it was about to begin. Under sustained shell-fire Jack galloped with his Regiment to seize a strategic objective, Tel Es Sakaty, and all around him he could see and hear thousands of mounted soldiers galloping to battle. Jack said he was scared of what was coming – he thought all men get scared at such times – but he wrote that there comes a sort of laughing courage from deep within the heart of each. He called it the terrible intoxication of war.[49] All morning Jack's troop was under fire until about midday they took Tel Es Sakaty. Next they were told to seize the Hebron to Beersheba Road because motor lorries carrying Turkish re-inforcements had been seen driving from the direction of Jerusalem.

Jack missed participating in the famous charge at Beersheba but from his observation post on a hill, through clouds of roiling dust, he tried to see what was happening. By the end of that day, Jack wrote, he and other observers could see the outer defences of Beersheba had been taken but not the city itself. Time was now short.

Just on sundown Jack said he saw regiment after regiment of horsemen advancing on Beersheba under intense fire: "*I heard shouts among the thundering hooves, saw balls of flame amongst those hooves – horse after horse crashed, but the massed squadrons thundered on*".[50] Then Jack saw

the artillery bursting behind the advancing lines of horsemen. This meant that the artillery gunners could not change their range fast enough. Suddenly men were not falling. Jack and his mates knew what this meant. The Turkish infantry in their excitement and fear had neglected to lower their sights. Bullets were flying over the heads of the horsemen.

So, with no artillery shells and fewer rifle bullets, Jack wrote that, "*the last half-mile* (nearly a kilometre) *was a berserk gallop with the squadrons in a magnificent line, a heart-throbbing sight as they plunged up the slope*". Jack saw horses leaping right over the defending trenches with Turks trying to bayonet the bellies of the flying horses. Then down another kilometre-long slope into the town itself. Jack wrote that from his hilltop he saw the last mass charge of mounted soldiers (this time against modern machine guns and artillery). He had witnessed the creation of a legend that resonates as strongly today as it did on that day in November 1917. "Beersheba had fallen".[51]

There are some analysts who dispute Jack's version in *The Desert Column* citing the geography of the area and differences between his diary and his book. For example, Jack's diary entry for 2 November 1917 does not record the events he described in his book and it has been suggested that he could not have seen the fall of Beersheba from his position.[52] The issue (as for any rewritten diary) is that brief diary entries are not in themselves book material. In writing a book from a diary there will be other unrecorded memories prompted by the diary entries. Jack's work is an uncommon instance where the original and the rewritten version are both available for comparison.[53] Naturally they are not always an exact match. For example in Jack's diaries, there was no mention of Simpson and his donkey but in *The Desert Column* Jack wrote, "… *The Infantry are quite cut up – not over their terrible losses, but because of one man, Simpson Kirkpatrick I think his name is…*".[54]

Whether or not Jack actually witnessed the charge at Beersheba, it was not the end of the Great War. Jack fought on. On 3 November he had his first experience of a new weapon – the howitzer from which a shell could drop almost vertically and so afford no cover from a cliff face. The shells were extraordinary. There was a double explosion that could take off a horse's head.[55] Even so, the Turks were on the run.

About a week later Jack's squadron was detailed to take two troublesome Turkish artillery positions. From a distance of about five kilometres the Brigade challenged the guns. Then Jack's Squadron, on its own, faced rifle and machine gun fire. There were lines of infantry behind each of the guns. Section after section raced at the guns, into a gully and over it then into another before they broke through the Turkish ranks and turned their rifles on the enemy. Just as they fixed their bayonets for a final attack on the retreating Turks the order came: "Retire!"[56]

Still in November, Jack participated in an exciting fighting retreat. His Squadron had got ahead of the Regiment and cantering over the ridge of a hill, they suddenly encountered a long line of about 1000 Turkish infantry. Vastly outnumbered (a Squadron is only about 100 men) Jack and his mates began an exhilarating game of life-and-death chance with the advancing Turkish foot soldiers. They lined the first ridge and fired on the Turks until they could see their eyes. Then, springing on to their waiting horses they galloped for a short distance. They swerved in behind a hill and again selected firing positions. Jack's Squadron could even hear the thump of the infantrymen's feet before they were again on their horses and galloping away. Again and again they did this (eventually with the help of another Squadron) and seriously delayed the infantry re-inforcements. When Jack's mob was joined by their whole Brigade the battle really began. The big force of Turks was annihilated.[57]

By this time Jack was sick with what was called "Jaffa Fever" – probably a strain of malaria. On 23 December 1917 he was on his back in the Australian General Hospital in Cairo. Not only did he have malaria but he also had the septic sores that (he said) four out of every five men suffered. The final blow for Jack came when he was on sick parade to have his sores dressed. He then acquired his third and most serious wound – the wound that finally put him out of action.

Sitting waiting for his sores to be dressed, Jack heard a shell coming. He threw himself on his face but it exploded too close. He wrote that he should have been blown to pieces but there was not even a bone broken. This description severely understated his injuries. He had twelve shell splinters in his body; some of which would stay with him for life.[58] So Jack had survived the Great War but at the age of twenty-eight years and four months, his days as a soldier were over.

Including their periods of hospitalisation, Jack was overseas for two years and eight months (18 May 1915 to 2 January 1918) while George was overseas for one year and four months (20 December 1917 to 19 April 1919). Exactly three years after Jack was discharged he and George were enjoying ten days of mateship on Howick Island.

29

Mud-pool
January 1921

The period of good fellowship in January 1921 mentioned in the second *Madman's Island* was disrupted over George's now-ripening sweet potatoes. Four and a half months earlier George had brought with him some cuttings or perhaps young tubers that were already shooting. Depending on climate, sweet potatoes can mature in two to ten months but around the Cooktown area they mature in about five months. Now, in the second week of January George's sweet potatoes were just about ready to harvest. Although Jack had consistently denigrated George's gardening, during those ten good days in January George had freely offered Jack some sweet potatoes. George added a condition. He asked Jack to pick them carefully and to replant the vines. Jack readily agreed to this condition.[1] Then one afternoon a few days later George went to his garden. What a shock. Jack had raided his garden and carelessly pulled the vines out by the roots. He had taken the tubers and just cast the vines aside – ruined.

Jack's account of his actions differs between the two books. In the first book, "*he hurriedly pulled handfuls of the vines up by the roots.*" In the second *Madman's Island*, "*he dug a lot and ... replanted the vines...*" In the first book Jack took so many tubers that he could hardly carry them all but in the second *Madman's Island*, he took "*a lot*" presumably just enough for his needs.[2]

The account in the first book is more likely to be correct because in the second *Madman's Island*, Jack does not explain why George (the "*grim wild figure*") suddenly arrived at the Mound to complain. Obviously, if Jack had taken up George's offer and carefully replanted the vines, George would not have had a reason to complain. From the account in the first book Jack would have had no doubt why George was angry – George was angry because Jack had ruined his garden.

There was good reason for George to be angry with Jack. The two men had been amicably talking for ten days. George had given Jack some wire for his spears and offered to share the sweet potatoes. Then Jack vandalised George's garden. George was furious and he set off across the reef to have it out with Jack.

It was late afternoon when George found Jack's handiwork in his vegetable garden.[3] Angrily, he hurried across the reef but he was taking a significant risk because the tide would soon be turning.[4] George did not care. He wanted to tell Jack what he thought about his behaviour. In his haste he walked right into a mud-pool and Jack wrote that he thought George would die.[5] Jack described the pool as, "*quick mud – slimy sticky stuff possessed of some horrible sucking power that could drag a man down*".[6] Jack described the mud-pool as a broad shimmering pool about fifteen metres from the base of the Mound and he had wondered why the family of cranes that fished in the pool never ventured to its centre.[7]

Just before sundown (as he finished his meal of George's sweet potatoes) Jack saw George approaching from across the reef. Jack saw George hesitate at the other side of the mud-pool. He then made a quick rush across the pool but as he reached the middle he instantly started floundering and sinking in the mud. Jack was in no doubt – he thought George would die. George screamed and Jack, "*breaking free of the icy chill of horror ... ran down the Mound*".[8] When he found himself sinking, George eventually decided to lie flat and use a swimming-type motion to extricate himself.[9]

George did find himself in quite a dangerous situation but it may not have been life-threatening. Peter Rutherford of Cooktown is a long-time resident of North Queensland and a keen Idriess reader. He found this incident to be one of the inconsistencies in *Madman's Island*. He said his daughter had found herself in a similar nasty predicament and had used the technique by which George used to work his way out of the mud. Peter said that with a cool head and slow careful movements there is no danger in these pools.[10]

But George did not have a cool head and time was short. George was very angry, it was getting dark and the tide was coming in. When he found himself in the mud he might have panicked and made his predicament worse than it should have been. More than likely, if he had not been in such an angry rush he probably would have avoided the mud-pool. Anyway, Jack saw him coming and he was a spectator to the whole incident and Jack's response to George's plight is one of the most curious parts of the whole Howick adventure.

George and Jack had just spent the first ten days of 1921 in easy companionship.[11] Jack had even considered shifting back to George's side of the island.[12] Back at the Mound (after raiding George's garden) when Jack first saw George coming he was wary. He probably expected George's retaliation. However, his first book Jack said he was, "*saturated with an ever-present instinct for self-preservation honed sharp after months of close danger* (from George)" and he wrote that he was suddenly on the alert when the cranes on the mud-pool took flight.[13]

Jack could see George was angry. Although he saw George had the rifle and Jack could be excused for assuming George intended to use it, the situation changed suddenly. George lost the rifle in the mud and Jack believed he was fighting for his life. Yet when Jack saw George struggling in the mud-pool he did not help.

As Peter Rutherford said, the escape method is to lie flat and with swimming motions work toward the edge of the mud.[14] After his initial panic George must have calmed himself and he did lie flat. He began to move his arms in a strong swimming motion toward the edge of the pool. He had another problem. The tide was coming in. The mud was becoming more liquid. George could just reach the tip of an overhanging branch. Slowly and carefully he pulled the twigs down until he could get hold of the branch; all the while continuing to make swimming strokes with the other arm.[15]

It was dark by the time George had grabbed the mangrove roots and reached safety. He was exhausted. Even so, he had to move quickly. As the tide rose he was in danger of drowning as he lay at the edge of the mud-pool. The water had already risen to fifteen centimetres deep around him.[16]

What was Jack doing as George struggled? The rifle was no longer an issue.

The rifle does not get a mention in Jack's books except when the incident was all over and Jack celebrated its loss in the mud-pool.[17] Jack was not in danger. It was George who was in urgent need of assistance.

In Jack's first book he wrote that he ran down the Mound but then said to himself, "*He's got a chance. Curse him; why should I help him anyway? Sooner or later he or I must die. It's survival of the fittest.*" As George reached the edge of the mud-pool, "*on the very point of shouting encouragement* (Jack, in the first book) *remained silent.*" Then when George did not move as the tide rose around him, Jack said to himself, "*Why doesn't he drag himself out? If he doesn't shift soon he'll drown. The tide will be over the reef in no time, and it is dark already. Curse him; why doesn't he make a good job of it one way or the other?*" In the first book Jack made no move to help George out of what Jack saw as a life-threatening dilemma.[18]

In the second *Madman's Island*, Jack's attitude was much more sympathetic. He said George was caught like a fly in a treacle pot when Jack jumped down from the Mound to throw George a stick.[19] Jack said the mud-pool was far too wide and long for him to have run around to help George. When George had reached safety and was lying at the edge of the pool (in the second *Madman's Island*) Jack shouted to George to hurry.[20]

However, in neither of his books did Jack do anything to help George. At best he shouted some encouragement. He was relieved when George survived. At worst he hoped George would die. But he did not help.

It was dark by the time George had grabbed the mangrove roots and pulled himself to safety. To get home he had the surf-lashed reef to cross. With the incoming tide, waves were breaking over the reef. The water was knee-deep and it was dark. He had a kilometre of swirling water to wade through before he could get to the safety of his camp. George was exhausted even before he started but before he finished that nightmare journey, he was staggering like a drunk.

Jack was watching from on top of the Mound. In the first book Jack wrote that he was "*greatly relieved*" when George survived the mud-pool but as George battled through the surf on the reef Jack, "*scowled vindictively*" and said to himself, "*May this little stunt teach you some sense, you murdering swine.*"[21] In the second *Madman's Island* Jack ran back up on the Mound to watch George's progress across the reef. In that book Jack said George was quite safe even though the water was up to his knees and he was crossing a coral reef in the dark.[22]

Then just two days later, Jack (at his most unpredictable) strolled across to George's camp. In the second *Madman's Island* when Jack turned up at George's camp he acted as though nothing had happened. Neither of them mentioned the mud-pool. Initially George was morose. He was probably still upset by Jack's raid on his vegetable garden. After a while he brightened and they talked amicably about George's idea of continuing to live on the island.

Of course, this period is not mentioned in Jack's first book. Indeed in that first book Jack said (referring to the loss of the rifle in the mud-pool), "…*if he tries to stop me* (getting water) *I shall fasten these two hands around his throat and press my knee into his stomach until there's not a gasp of wind in his whole cursed body.*"[23] In his first book Jack reported little conversation between himself and George.

Half-way through January 1921 George and Jack were again talking amicably (despite Jack's raid on George's vegetable garden and the mud-pool incident) but Jack provides no indication of the content of their discussions. Nevertheless, at some time the two men would have talked about their homecoming and their lives after the war.

30

After the War

After the war George was not well. He had arrived back in Australia with what Jack described as a fearful scar and a gruesome looking hole in his abdomen. His doctors reported that George always felt low-spirited and unwell.[1] George was unable to walk uphill without breathlessness but (deliberately disregarding his disabilities) he promptly got back to mining in North Queensland.

George was discharged on 19 April 1919 and by October he was in Cairns. He arrived on the *Bombala* on October 11[2] and on 31 October George and four other ex-servicemen received a welcome home.[3]

George might have stopped off at Cairns to see his brother William. William had also been to war and he had returned to Australia on 11 December 1918.[4] After arriving in Australia with George and their brother Washington in 1888, William lived all his life around the Cairns area. After the war he lived at Tully River and Innisfail. Eventually he was admitted as a long-term resident of the Eventide Home at Innisfail. William died on 17 February 1937.[5]

However, a stop-off in Cairns was only a short diversion from George's intended objective on the Cape York Peninsula. Within a month or so George was back at work. In the same year (1919) that he was in Cairns in October, George was back at Spion Kop, Yarraden as a miner. So ill-health notwithstanding, George got straight back into what he knew best – chasing gold.[6]

Jack was also a sick man when he arrived back in Australia. He returned to Australia on the *Ulysses* on 15 February 1918. He was discharged with several pieces of shrapnel still lodged in his body and he was told he would never again walk without crutches. Beverley Eley said doctors in Cairo had been forced to leave a large piece of shrapnel lodged against his spinal column, another large piece in his right thigh and numerous smaller pieces elsewhere in his body.[7] Unfortunately Eley's biography does not fully explain Jack's life after the war.

Although Beverley Eley revealed little of what Jack eventually did after the war, it is true that he attended the gymnasium of Snowy Baker for a short time. Baker's biographer billed him as "*Australia's greatest all-round sportsman,* (who) *excelled in 26 different sports*". Jack got a brief inaccurate mention in that biography[9] and he briefly toured with Baker's travelling boxing troupe. In August 1918 at Grafton NSW he fought a three-round exhibition bout with an American World Champion lightweight boxer.[10] Then by March the next year (obviously on his way back north to Cooktown), Jack was in Brisbane where he witnessed a violent trade unionist demonstration.[11]

Jack did not waste time around Brisbane. Sometime well prior to October 1919 he was back prospecting with three mates. On 18 October 1919 the *Brisbane Courier* reported:

> "*The Queensland Government Mining Journal for October 13 contains... an interesting report by J. L. Idriess and party (returned soldiers), on prospecting work between the heads of the Normanby and Daintree Rivers. Work is carried on in rough country with dense scrub, and a long way from sources of provisions, etc. The men have been working mainly with sluicing operations, and have seen some fairly encouraging prospects with gold in pieces up to a 7dwt. piece.*"[12]

It is likely that Jack was continuously in the same area for the next five months. On 16 March 1920 the *Cairns Post* reported:

> "*Idriess and party of three returned soldiers, report having prospected for gold on the Daintree and Normanby Rivers for some months past with little or no success. Practically the whole period they obtained 16 ounces of coarse gold. They have decided, owing to the excessive wet season and the extremely mountainous and dense scrub country in which they have been working, to abandon prospecting*

until the wet season is over, and to follow prospecting for tin and around the tinfields. (They have been assisted by the Department by means of rations, etc.)."[13]

It is not possible to nominate with any certainty the three returned soldiers who were working with Jack, but surely Jack would have been anxious to renew his friendships with Norman and Charlie Baird and with Dick Welsh. Eley said that in this period Jack and Dick rode to the tip of the Cape York Peninsula but (given the press reports above) this is not likely to be correct.

In any case when Dick Welsh returned to Australia he was in bad shape. Eley said he was a "*morose alcoholic*".[14] Although he would have tried to get back into prospecting and mining (and he could have been one of the three ex-servicemen working with Jack), Eley's depiction of the post-war and pre-Howick exploring and prospecting trip along the whole length of the Peninsula (right to the tip of Cape York)[15] could not have been managed in the time available. From another source it was said that Dick was not well. Jim McJannett, Cooktown resident and researcher into local history, said Dick worked for a while at the Irvineville Tin Mine but he could not manage the work.[16]

One of Jack's mates who was definitely not in the Idriess mining party was Charlie Baird. Charlie served with the 11th Light Horse Regiment Cooktown[17] but when he was discharged he never returned to the Cooktown area.[18]

It is quite possible that Norman Baird was one of Jack's party. Norman was one of the most remarkable men Jack met in his travels. The son of a Scottish father and an Aboriginal mother he had a strong sense of civic duty and a firm grounding in both European and Aboriginal culture. Intelligent, well-read, opinionated and outspoken he was fluent in both languages.

When Norman returned home from the First World War he was not treated like other ex-servicemen. Because of his outspoken advocacy for his Kuku Yalanji people he was constantly in danger of being removed from the Bloomfield area to Palm Island.[19] Norman worked tirelessly and forcefully to resist the relocation of his family. He used the knowledge from his European education combined with the skills and bushmanship of his Aboriginal ancestors to frustrate and evade the Queensland Protector of Aborigines who wanted to exile him to Palm Island. Nevertheless his son, Joseph was relocated to Cairns and Norman did not see him again for ten years. Norman and his descendants continued to live in Kuku Yalanji and Guugu Yimithirr country.[20]

Yarning by George's campfire after George had survived the mud-pool, George and Jack might have talked about their post-war experiences and perhaps they shared their successes and disappointments in chasing gold. Now they were about to find some smuggled opium.

31

Opium
January 1921

Along with the romantic interest (that makes up well over half the first book), it is the story of Jack finding some smuggled opium that is one of the main differences between the first book and the second *Madman's Island.* The basic story of the opium is similar in both books and Jack has recounted the same story in his books *The Opium Smugglers* and *The Yellow Joss.* He also wrote magazine articles about the incident.[1]

A buoy containing opium was thrown overboard from a legitimate vessel that was passing a pre-arranged remote location on the North Queensland coast. It was to have been picked up by smugglers in a smaller boat posing as a fishing vessel. To help the smugglers find the buoy it had a small flagpole on which there was a light and a flag. For some reason the smugglers were late. Jack found the buoy and the smugglers lost their opium.

George and Jack would have been well aware of opium because of their knowledge of Cooktown and the occasional press report of opium raids by the police on Chinatown. Social attitudes toward the addictive properties of opium and other narcotics had been hardening from around the 1890s. In Australian States the possession and smoking of opium was prohibited from around 1906-1908. The fine for possessing opium was about £50 with six months hard labour in gaol if the fine was not paid. However, even up until the early part of the twentieth century opium was still the main ingredient in products like paregoric and laudanum which were openly sold to calm the nerves and relieve pain.

About half-way through January 1921, Jack found the smuggled opium after a large passenger vessel (Jack called it a "China boat") with lights ablaze and music playing passed by in the main shipping channel. Next morning Jack found a sealed tin of opium that had been thrown overboard for smugglers to retrieve. It is what Jack said happened next that helps to differentiate the two *Madman's Island* books.

In the first book Jack wrote that he took the tin to his cave and broke it open to find 144 tins of opium worth about £20 per tin.[2] Then in the first book, Jack (calling himself "Jack Burnet") took the opium with him when he left the island just over a week later.[3] Jack wrote that "Jack Burnet" sold half of the opium. After adventures with the police and the smugglers[4] "Jack Burnet" used the money from the opium to get married and set himself up as a farmer.[5]

Of course this is not true. Jack never did get married and he never became a farmer but the mystery is not resolved in the second *Madman's Island*. Jack provides no other explanation of what really did happen to the opium.

There on a tiny isolated island upon which they have been stranded for four-and-a-half months Jack wrote that he and George suddenly found themselves in possession of an illegal substance worth (Jack said) nearly £3000. By comparison, he thought the opium would be worth thirty times more than he and George expected to get for the tin and wolfram they had so laboriously dug and bagged from the Peak on Howick Island.

In the first book George was not told about the opium ("Jack Burnet" kept it for himself) but in the second *Madman's Island* as soon as he could, Jack splashed across the nearly dry reef to show it to George and they opened the buoy together.[6] Jack gave no indication that they talked about disposing of the opium. Nor did they talk about smoking it. Even though he said they "*were dying for a smoke* (of tobacco)" and despite knowing the value of the opium, Jack wrote that "*the beastly stuff*" was worthless to them.[7] Although Jack did not say so, they would have talked about their options but first they had to deal with the smugglers standing off-shore in their sinister black cutter. In three dinghies, the smugglers began searching the mangroves.[8]

Jack's first thought was to give up the opium in return for a ride back to

Cooktown. George emphatically told him that an approach to the smugglers would earn them bullets in their heads. They set about hiding any sign of their presence. Jack's signal flag on the Peak came down. Initially they were worried the smugglers might have seen the flag but they had arrived at night and by morning were anchored too close in to shore to see the flag on top of the Peak.[9]

George and Jack banked down their cooking fire and made sure nothing was lying about that might show there was anyone living on the island. The smugglers could not see George's camp and they would not be interested in the Mound. For a whole week the smugglers searched the mangrove roots along the shore of Howick Island, Houghton Island and Coquet Island. Hiding on the Peak, George and Jack watched them. It was a tense and dangerous time because at any time the smugglers could have decided to come inland and look at the island more closely.

Jack could not get back across the reef for fear of being seen. His fire would have gone out but they kept the coals of George's fire alive and carefully cooked crabs and the occasional fish. George's experience living with Aboriginal people had taught him to make small almost undetectable cooking fires.

Now, for the first time in nearly five months George and Jack were together facing a deadly enemy. United in their resolve to survive a possible attack from the smugglers, that week was better for their relationship than all the rest of the time they had been on the island.[10]

They made all sorts of contingency plans. What if the smugglers knew about the well and found George's camp? What if the smugglers spotted either of them? They had intense quiet debate about those possibilities. They even considered a counter-attack.[11]

George and Jack decided that if the smugglers tried to get them they would have to split up to guard the Peak, the Hill, the Mound, the well and also their dinghies and their cutter. George and Jack thought this would stretch the resources of the smugglers and that they would be attacking only one or two men at a time. What if the smugglers left the cutter unguarded? They considered seizing the cutter and marooning the smugglers. Of course this

stuff was much too fanciful. The smugglers had guns and George and Jack had only a couple of little fishing spears. Even so, debating their defence drew them together in a way that they had not been close before.

This went on for a whole tense week then finally one day, just on sunset, the cutter's sails filled and it was away on some other nefarious journey. George and Jack were left with 144 tins of opium. They could have opened the tins and thrown the opium into the sea. They could have taken it back to Cooktown but the consequences of that action were quite easy to predict. If they tried to sell it they could easily be in trouble with the Police and/or the smugglers (just as described in Jack's first book). If they handed it in to the Cooktown police the smugglers could at any future time, get their revenge in any number of ways.

They might even have talked about smoking some of it. These two confirmed tobacco smokers had been out of tobacco for months and there was no prospect of rescue in sight. The temptation must have been strong but both would have had first-hand knowledge of Cooktown's Chinatown and would have seen the effect of opium on its addicts. It must also be said that both men had a problem with alcohol dependence.

Whatever did happen; the opium just sat there by the campfire in 144 evil little tins.

George and Jack did not yet know it but their time together was almost at its end. For the last week in January and the first week of February they were the best of mates. George even asked Jack to move back to his camp but Jack declined.[12] For the first time since they arrived on the island they were talking not just about their previous experiences but their thoughts and feelings.[13]

32

Thoughts and Feelings

Right from the start of both the *Madman's Island* books, George was talking about living on the island. In the first book (because the occasions during which the two men conversed were so few) George only voiced his opinions on two occasions. In their third week on the island, after "*a fortnight of bright days, gentle breezes, and cool, quiet nights* (had) *drifted calmly by*".[1] George talked with Jack about his ideas. Jack was incredulous when George told him he wanted to live on the island. George told him:

> "*Why not? A man could make a good garden, and have vegetables all the year round. He could get a few fowls from Cooktown and let them breed. There's abundance of fish. All a man would want would be to arrange with a cutter to call every twelve months with a few bags of flour, tea, sugar and tobacco–and he could do without that at a pinch*"[2]

Jack argued with him and asked George about his sense of companionship, about seeing and talking with other people. George was adamant:

> "*You can keep your hotbeds of disease and squalor, your wretched picture-shows and artificial people. What has the world done for me anyway? Nothing but bullocking graft to earn a crust, and starve if there was no graft to get! And all your 'human companionship' thinks of is to take the other fellow down if he's got money, and kick him into the gutter if he hasn't. The world's got no time for me and I've got no time for the world. Blast the world! They can cut their brother's throats and see their sisters tread the gutter for all I care! The world can go its own way to hell for my part*".[3]

In the first book the only other reference to George's attitude was when, just as Jack was about to be rescued George told Jack, "*I'm never going to leave the island. I'll live and die here.*"[4]

After Jack was rescued he made a statement to the Cooktown police. In that statement he referred to George's attitude. Jack was reported as saying, "*He said he might just as well die on the island as in the Cooktown Hospital. He added that civilisation had never done anything for him, and he was sick of the rotten ways of his fellow men.*"[5]

In the second *Madman's Island* Jack wrote more fully about George's attitudes. Even as they enjoyed their first meal on the island George told Jack there was enough driftwood to build a comfortable hut.[6] The conversation described in the first book (when George referred to "hotbeds of disease and squalor") was also described in the second *Madman's Island* but later than described in the first book – about five weeks after they had arrived.[7] When Moynahan's ketch finally did arrive (but then sailed away without seeing Jack's frantic signals), Jack suggested to George that he should get ready to leave. George grumpily refused.[8]

As Jack fished the mangrove creeks, George tended his garden. Jack called him a born gardener. George still held to the idea that he could live on the island. He thought there might still be tin and wolfram to be won from the Peak. He thought that if he had a dinghy he could cruise from island to island and across to the mainland.[9] He even (fancifully) discussed attacking the opium smugglers and seizing their boat.[10]

George's idea was that the minerals he believed he could get from the island would help him buy tools and build a comfortable hut. With the money he could earn from mining he could keep the chemicals he needed for his bowels. He would then (as Jack wrote), "*be absolutely independent of the world, his own master in every respect for the remainder of his life*".[11]

Jack never could understand these dreams. He wondered how George could live so alone and not be afraid of his own thoughts. It puzzled Jack. He could not believe a man could voluntarily resign himself to a totally solitary life.[12]

Although neither man knew it, at the end of January 1921 George's ideas were about to be tested. Jack's rescue was close and George was about to be left alone on Howick Island. Ironically, by this time the two men had developed a relationship in which they were able to talk freely not just about their life experiences but also about their thoughts and feelings. Of course, in his book Jack is putting words into George's mouth but the views expressed are not the same as Jack's ideas. They could reasonably be attributable to George.

George explained to Jack his sense of the presence of "Me" that he believed was in him. He could have been talking more-or-less about his soul but George felt that in some way the "Me" survives the death of the body but retains consciousness. He said when "*his damned cumbersome body*" with all its health problems that "*keep a man chained like an animal to a cart*" dies, the Me keeps on living. Then George was certain that when his body dies the Me would be freed.[13] In some way he seemed to feel that he would still have consciousness and control but without his body to restrict him. In talking with Jack he seemed to be saying his sense of Me separated him from the world around him and at once made him an individual but also part of something much bigger.

Whilst in some respects Jack had somewhat similar views to those he attributed to George, he also believed the living could contact the dead.[14] Eley wrote about Jack's belief in mysticism, mental telepathy and other psychic phenomena. Eley said: "*Many times he had seen the Aborigines respond to warnings from their brothers, warnings which came unbidden, and many times saved them by death at the hands of enemy tribes.*"[15]

Jack referred to himself as a mystic. He believed living people could commune with people who had died (or in another sense, passed on from this world to another). This comes through in some of Jack's books. Eley mentioned Mister Limdley (who edited *The Yellow Joss*) that, "*he only cut out one little piece of supernatural business*" while editing the book.[16]

Jack's enduring belief in Spiritualism affords no real clue toward his attitude to Christianity and Christians (although the beliefs of Spiritualism seem antithetic to the beliefs of Christianity). While from Jack's books there is no obvious clue to his attitude toward religious beliefs, he certainly did maintain many personal relationships with men of religion and in *Drums of Mer* he had his main character discuss concepts of Christianity at length.[17]

In his wartime diaries (long before he was talking with George about life after death), Jack had written of a type of supernatural experience. Jack said he was on night-time sentry duty in the desert. He began to think of all the different armies that had fought over this land. He imagined the spirits of strange phantom soldiers standing over the bodies of dying Anzacs and he, "*...could see quite plainly the spirits of Anzacs arising and staring at the weird soldiers gazing so silently back.*" The experience became so real that Jack forgot his sentry duties. He said a phantom came to stand beside him; "*...a calm, steady sort of chap standing beside me explaining, not in rough speech, but putting pictures in my mind...*" to show Jack the huge armies of long-dead soldiers watching over his current war and (just in time) to remind him of his earthly sentry duties.[18]

Just about a week or so later Jack wrote about a ghostly horde of horsemen in the desert. To his disgust his unit was left out of a major battle at Maghdaba. In *The Desert Column* Jack quoted his diary entry for Christmas Day 1916 in which he described the battle he had missed then his diary continued:

> "*Later–A very peculiar story is being discussed throughout the Desert Column. It appears that the troops when riding back the thirty miles from Maghdaba were enveloped in blinding clouds of dust. Nearly the whole column was riding in snatches of sleep; no one had slept for four nights and they had ridden ninety miles. Hundreds of men saw the queerest visions–weird looking soldiers were riding beside them, many were mounted on strange animals. Hordes walked right amongst the horses making not the slightest sound. The column rode through towns with lights gleaming from the shuttered windows of quaint buildings. The country was all waving green fields, and trees and flower gardens. Numbers of the men are speaking of what they saw in a most interesting, queer way. There were tall stone temples with marble pillars and swinging oil lamps–our fellows could smell the incense–and white mosques with stately minarets.*"[19]

Over two decades later when Jack was researching material in Broome

(Western Australia) he tried to make contact with ghosts by spending the night in a graveyard and on a reputedly haunted verandah.[20]

George's belief in the "Me" inside his earthly body was quite different from Jack's beliefs. Only a part of Spiritualism is concerned with life after death. Jack's Spiritualism departed from George's beliefs (and the beliefs of Christians) in that it also incorporated, "*...the demonstrated fact of communication, by means of medium-ship, with those who live in the spirit world.*"[21] Jack believed he could use the gifts of a living medium to communicate with the spirits of people who had died.[22] Jack was also interested in mysticism, mental telepathy and other psychic phenomena. Eley said some people (including Professor Colin Roderick) had referred to Jack as a mystic.[23]

In the Eley biography there is also mention of Jack delving in to the occult over a long period of time. She said Jack and his friends had held occasional séances in the 1930s trying to contact Harold Bell Lasseter who had disappeared after reporting finding a fabulously rich gold deposit.[24]

Eley said Jack was still studying the occult in 1933 and quoted a letter responding to a request for information about a medium. Jack's response was, "*I am quite convinced of survival after two years of study of various phenomena. But it is extraordinarily hard to come by a medium who can produce anything that counts.*"[25] Much later, when Jack was about eighty years of age, he wrote to Alma Timms (widow of the author E.V. Timms) to say Timms had contacted him after his death. Jack said Timms had urged Alma to complete the unfinished manuscript of his last book, *Big Country*. Alma, writing back to Jack said she also had an affinity with the "beyond". She wrote, "*There IS something beyond*" and she did complete and publish the book.[26]

On Howick Island, as January 1921 was coming to a close, Jack wrote that he talked with George about life after death. George was sure there is life after death. he said the "Me" in him was only there for the time being.[27] Jack said George had never spoken like this even in the good days of their first month.[28]

33

Rescue
February 1921

Some time early in February 1921 Jack was rescued from Howick Island. His rescue was dangerous because of the rough sea breaking over the reef. From the top of the Mound where he could look out over the mangrove tops to Cape Bedford, all day Jack watched his rescuers approaching the island. This time they did not sail away. His signalling was seen and Jack was taken back to Cooktown. George would not leave. In both of Jack's *Madman's Island* books, Jack tried hard to persuade George to go with him but he stubbornly refused to leave the island. In the second *Madman's Island*, however, Jack said they parted on good terms.

After the mud-pool incident (and after the smugglers had given up their search for their opium), during the last week in January and the first week of February George and Jack were the best of mates. Jack said they "*were good friends*" and George was "*... the best of companions.*"[1] Quite soon after Jack found the opium the weather turned from calm to storm. In fact the weather worsened while Jack was almost about to be rescued. He had seen the sails of pearling boats heading for Howick and he hurried to the Peak to build up an already existing cairn. He had a bamboo flagpole and the remains of his blanket for a distress flag. He worked all morning on the cairn and was surprised at midday to find the wind had grown to powerful proportions. Jack also tried to get George to leave.

In the first book Jack approached George "*freely but warily*" to talk to him about leaving. This was the first word they had spoken since early December when they had their second fight, that is, when George found lice in his blanket the two men fought and George fired the rifle at Jack. In the first book the two men did not speak again until rescue was imminent. Jack thought George might attack him but he was confident he could defend himself. As it turned out he found George to be "*sane again*... (but)... *horribly woebegone*." Jack "*pityingly*" tried without success to persuade George to leave.[2]

There were many differences in the way the two *Madman's Island* books were written. Of course the most obvious difference is the romantic part of the first book that takes up 127 of its 228 pages. The mystery of the opium is another difference but the main distinction between the first book and the second *Madman's Island* is in the way that Jack portrayed his relationship with George. Nowhere is this distinction more pronounced than in their parting conversation.

In the second *Madman's Island*, Jack had two parting conversations with George – in the morning as the rescue boats first appeared and again in the evening as they neared the island. Getting toward sundown the two men had a long conversation and far from being "mad", Jack wrote that George "... *was quite in his right mind*."[3] During this conversation they officially transferred and signed over ownership of their right to mine the island and Jack gave to George his share of the tin and wolfram they had bagged. Although George (Jack wrote) was "*morose*" and "*stubborn*" they shook hands and parted on good terms.[4]

As the pearling boats came closer the crew fired a gun to show Jack they had seen him but he had to wait overnight while they sheltered behind Coquet Island. Next morning Jack was rescued by a dinghy from one of the pearlers. The conditions could not have been more different to those of his arrival. There were high winds and strong waves. The dinghy could not get across the reef so Jack had to push through the waves out to the dinghy.

Heavy waves were rolling in toward the island, smashing on the reef with a thunderous crash. A powerful wind had whipped the sea into a swirling lather of foam. Jack faced a huge challenge to get through the surf to a dinghy lowered from one of the pearling boats. Despite being knocked down by the power of the waves, he leapt up and plunged out again. It was Jack's desperation that conquered his fear of the danger.[5]

Jack had survived nearly five months on a tiny island living off the resources of the sea. The men in the dinghy lifted him to safety and he was finally rescued.

Part Three

What happened next and the truth about Charlie.

34

George's "madness"

In February 1921 Jack had left George on Howick Island. By 1928, at only 54 years of age, George was dead but some seven months earlier Jack had nominated him as the madman of the first *Madman's Island* book. After reviewing his story, the question remains as to whether George was really "mad" as Jack claimed.

It is for the readers of the two *Madman's Island* books to decide for themselves whether George was "mad" but (even given the problems of making a diagnosis some ninety years later) it seems unlikely that George had a psychiatric illness. It is far more likely that (for the purposes of his story in *Madman's Island*) Jack exaggerated George's taciturn personality and his emotional state.

George did not have to be mad for the relationship between himself and Jack to break down. These two men should never have gone to Howick Island. Only a few years earlier, both men had returned – damaged – from the First World War. Both were "loners" in different ways. Jack was happy to be with other people but he had deliberately acquired the ability to keep himself separate from them. George lived and worked among other people but he did not "need" them.

More importantly, when they set out on the prospecting adventure in September 1920 they knew almost nothing about each other. They might have been on one brief prospecting trip while they were waiting for transport to Howick Island. Even though they intended staying for only a month, Howick Island would certainly not be the place to learn for the first time about another person's character. A breakdown in their relationship was absolutely predictable. And Jack would not have been the ideal companion.

Even in their first hour or two on the island, Jack was away daydreaming as George set up the camp. George was not happy but Jack just laughed off his implied reproach. And another potential problem emerged right at the outset. George ran a tidy and efficient camp – Jack called it "shipshape". On the other hand Jack freely admitted that his camp housekeeping was carelessly untidy. Jack emphasised this personality trait in *Madman's Island* and other books. Right from their first few days on the island George must have started to get irritated with Jack's apparent laziness. It is quite likely that the seeds of discord were already being sown.

However, there were incidents that were much more serious and not easily explained away. George did shoot at Jack – twice. The second time (after Jack had tried to kill him with rocks) he had no hope of hitting Jack from across the reef and he would have known so. It was an expression of sheer frustration. However, on the first occasion he did shoot directly at Jack. The bullets hit trees close enough for Jack to hear them. Jack believed this action was a clear indication of "madness".

Then another unexplainable incident occurred after Jack was safe on the Mound. Because he had no water (or food or fire) he decided to raid George's camp. He found George guarding the only source of water on the island. George's reasoning in guarding the well is unclear and again not easily explained away. He would have known of Jack's predicament. He left his camp unguarded and patrolled the well – at least to prevent Jack having access to the water. Jack thought George's plan was to ambush and shoot him.

George's behaviour in keeping Jack from the well might indicate more than mere frustration. There was a good deal of vindictiveness in this action. Similarly, after one of Jack's raids on George's camp, George reacted spitefully. He found Jack's store of water and filled the water cans with sand.

Even so, after that time every one of George's aggressive acts was a reaction to Jack's raids. These raids were not benign. On Jack's second raid he burned George's tent. Then Jack raided George's vegetable garden. This was the very garden that Jack had originally mocked. After each raid George

retaliated. So, except for the two incidents where George shot at Jack and later when he decided to lie in wait for Jack at the well, all George's reactions were understandable in terms of two dissimilar men falling out in difficult circumstances.

It can reasonably be argued that Jack unfairly categorised George as a "madman". He described their relationship quite differently in his two *Madman's Island* books. In the first book their relationship began deteriorating about the time of the sand-flies and worsened. After they had two fist fights the relationship turned to irreconcilable and bitter enmity. In the first book the two men did not speak again until Jack was about to be rescued. Only when rescue was at hand did Jack warily approach George to tell him about the opportunity for rescue.

Their relationship was presented as much more complex in the second *Madman's Island.* In their first two months their relationship was reasonable although tension was building. Then in November the relationship worsened. In December there was a period of open hostility. In January their relationship improved and before Jack was rescued in February the two men were not only on speaking terms – they were revealing their emotions and thoughts.

In the second *Madman's Island*, Jack wrote that George drifted in and out of madness depending on his need for irrigating his bowel. Jack's attitude toward George was softened in the second *Madman's Island.* For example, in the first book, after being told by George to do his own fishing when Jack eventually managed to spear some crabs he did not offer to share; indeed he resolved not to tell George about the crab burrows, telling himself that George had turned out to be a rotten mate. In the second *Madman's Island*, Jack was much more conciliatory. When Jack caught some crabs he offered George a share.

It was early in the second *Madman's Island* that Jack laid the foundations for his allegation of George's madness. It is likely that Jack exaggerated what he saw as irrational behaviour in George. Certainly Jack made much of George's moodiness. Jack described George as morose, sombre and looking positively savage. He had George growling rather than speaking. Jack said he had been warned (even before they left Cooktown) that George would sometimes go off his head. In a back-handed concession, however, Jack wrote that George seemed to be as sane as himself. Of course, that's all from Jack's perspective – he was writing the story.

So, even from Jack's description of George's behaviour, it does not seem likely that George was living with a diagnosable psychiatric disorder. However, he might have been living with depression. In the second *Madman's Island*, Jack had George describe his "humours". George told Jack that when something steals over him his thoughts would be as black as hell. Depending on these humours, George said that at other times he could be constructively thoughtful or so cheerful that he could whistle instead of think.[1] George told Jack that he might be "as right as pie" but the humours swayed him against his will. He said that at times his thoughts would be quite all right (and he always tried to steer them straight) but the humours would sometimes prevent him having control over his thoughts.[2]

It is reasonable to accept these thoughts about "humours" (even though Jack is writing the story and putting words in George's mouth) and it is also reasonable (with some allowances for literary licence) to accept Jack's earlier depictions of George's personality and moodiness. These could be descriptors of depression and the symptoms of depression cover some of George's behaviour (as reported by Jack).

Depression is a syndrome of symptoms that can (among other symptoms) include (a) social withdrawal and a tendency toward isolation, (b) irritability, (c) fluctuations in mood and (d) sudden anger for no good reason. Admittedly, these symptoms have been "cherry-picked" to fit with Jack's description of George but a clinically depressed person will also often have serious problems interacting with other people. Not infrequently they blame other people or events for their feelings. Depression might have exacerbated George's irritability with Jack's laziness.

A person living with depression can also experience anxiety when around groups of people and even come to avoid contact with other people whenever they can. Nevertheless, this tendency to isolation would not necessarily preclude (as George did) working behind a hotel bar but George did clearly describe his preference for his own company. This might have been a symptom of depression.

Significantly, clinical depression is cyclical. A person living with depression has "good" and "bad" days – even bad mornings that might improve during the day. This fits with one of the main differences between the two *Madman's Island* books. In the first book George was continuously "mad" but in the second book his condition frequently fluctuated.

The symptoms of depression selected in the previous paragraphs can also overlap with Bi-polar Disorder or Schizophrenia. However, Jack did not describe any of the specific symptoms of these conditions. Opposing Jack's assertion that George was "mad", there are several contrary indicators.

Firstly, when Constable Brown went to Howick to check on George, his brief would have included verifying Jack's opinion that George was not "*right in the head*". Yet when he returned to Cooktown Brown seemed satisfied that George was managing adequately. There was no mention of madness and Brown left George on the island by himself.[3]

Secondly, Jim McJannett (Cooktown resident and researcher of the lives of Cape York pioneers) said he talked with Joseph George (Dick) Hatfield who knew both Jack and George. Admittedly hearsay, McJannett said Hatfield's words were, "*Trit was not deranged before he went to Howick or after it*".[4]

Thirdly, subsequent to the Howick adventure, George was working with other miners on gold claims. Mining Wardens' Reports lead to the conclusion that George was the principal of at least some of these ventures. This certainly does not fit with the idea that George was living with a psychiatric disability.[5]

A justifiable conclusion from reading Jack's accounts (in the two *Madman's Island* books) is that George was not "mad" but it is possible he was living with depression.

35

Jack's Emotional State

Of course, when people have been dead for decades it is more than a little presumptuous to offer even tentative diagnoses about them but for George, a diagnosis of depression would at least partially explain his behaviour. With the same reservations, it is also quite possible that Jack was himself living with an anxiety disorder originally connected to the death of his mother and exacerbated by his war experiences. With the reservation that a diagnosis at this distance in time is presumptuous, a specific anxiety condition (today called Post-traumatic Stress Disorder or PTSD) would explain a lot of the darker side of Jack's later life.

PTSD is recognised, classified and defined by mental health professionals and by the manual of disorders produced by the American Psychiatric Association[1] as an anxiety disorder. However, PTSD is not a mental illness as such; it is a response to extreme emotional trauma that has been observed in all cultures. Thousands of soldiers before and since The Great War have been left with this condition.

Although the name PTSD is fairly new, the symptoms that make up the condition have long been recognised. These symptoms were described during the American Civil War in the 1860s when combat veterans were referred to as suffering from "soldier's heart."[2] In World War I, the symptoms of the condition now known as PTSD were referred to as "combat fatigue". Later terms for the same symptoms were "war neurosis", "shell shock" "gross stress reaction", and "post-Vietnam syndrome". Even as far back as 1666 following the Great Fire of London, the diarist Samuel Pepys described symptoms that today would be accepted as symptoms of PTSD.[3]

It is possible that the circumstances surrounding the death of Jack's mother could have set him up for the PTSD that later developed as a result of his war service. In November 1907 Jack contracted typhoid fever and his mother nursed him at home until he was rushed to hospital. Jack never saw his mother again – she died of typhoid in January 1908. Soon after he was discharged from hospital Jack was left in Sydney by is father. For the rest of his life Jack had a deep sense of guilt over his mother's death and by abruptly sending Jack away from his family, Jack's father might have unwittingly re-inforced this sense of blame.

Studies by Breslau *et al* (1995) and Emery *et al* (1991) both connected severe negative childhood events (or prior exposure to trauma) with PTSD in ex-servicemen.[4] The circumstances of Jack's mother's death and his sudden separation from his family would have had an adverse effect on Jack's personality. These circumstances could have pre-disposed him to the symptoms of PTSD. After his mother's death Jack cut himself off from other people. In Beverley Eley's opinion, he deliberately perfected the art of saying little to those around him. Jack seemed to have built a personality based on a sort of separate existence; a personality that could come across as not so much self-possessed as insular or even selfish.[5]

Probably one of the most graphic descriptions of Jack's capacity to isolate himself from others was given by Gus Gaunt, a war-time comrade who said that to even his closest associates Jack was an enigma and to get a word out of him was almost as great a problem as to fathom his thoughts. Gaunt said he was the silent listener to the arguments of his fellow diggers and only a "*flicker of amusement might pass across his mostly sphinx like face*" but he would not join in with his own opinions.[6]

Throughout his life Jack gave to the world almost nothing away about his personality. In a 1963 publication, many prominent writers were invited to write about themselves and most of them provided first-person insights into their personality. Jack wrote nothing about himself. His profile is simply a chronicle of his achievements.[7]

Eley said Jack cultivated a rigid control over his emotions but it is possible that what she saw as deliberately cultivated "separateness" might have been a symptom of PTSD. The feeling of detachment or estrangement from others

is one of the key symptoms of PTSD. Similarly, Eley's view of Jack as being deliberately unwilling to express or share his emotions might have been an inability to do so. People who live with this disorder describe this symptom as emotional numbness.

Although Jack developed his own strategies for coping with his distressing negative thoughts about his mother's death, it is far more likely he developed PTSD from his war experiences. Hypothetically, if today someone with Jack's wartime experiences applied for a pension under the Statements of Principles laid down by the Australian Repatriation Medical Authority, there would be no doubt at all of his eligibility.

In this hypothetical approach to the Repatriation Commission, Jack would claim serious injury to himself. He was nearly killed. He would relate witnessing hundreds of deaths and living with the sight and smell of decaying corpses. He would tell the Repatriation Commission about the people he himself had killed. As he wrote in his diaries (and eventually in *The Desert Column*) he described his fear and horror.[8] There is no doubt that today Jack would be eligible for a Totally and Permanently Incapacitated disability pension on the grounds of PTSD.

In Beverley Eley's biography she describes other aspects of Jack's personality that point to the conclusion that he was living with PTSD. People with PTSD frequently adopt negative coping strategies. Jack seems to have had several. Isolation has already been mentioned but depression, anger and irritability also seem to have been a major part of Jack's later life.[9]

His chosen career of writing and researching had become his own private hiding place. No other profession (with the exception perhaps of a long-distance swimmer or runner) provides and requires such isolation. Living in his head, in his own world, and avoiding emotional attachment he had managed to observe the world with tolerant, even-handed, even amused detachment. He could be passionate about Australia, Aborigines, Australian authors and Australian publishing, but never passionate about individual people.

Addictive behaviour is also common in people with PTSD and Eley describes Jack's smoking, gambling and heavy drinking. There is no evidence

that Jack used drugs but he did mention opium in several books including the mystery of the opium in both of the *Madman's Island* books.

Beverley Eley chronicles Jack's later life (not covered in detail by this book) and said his later life revolved around avoiding his home and the family. He achieved this by working at Angus and Robertson and then drinking for hours with his mates. Even when he finally did come home, Eley said Jack would shut himself up in his room to think and write for hours on end.[10]

Whatever problems lay between Jack and Eta (Jack's common-law wife) they could never be resolved by Jack's sparse, uncommunicative style of relating to other people. He had consciously or unconsciously developed an introspective personality and his increasing dependence on alcohol seemed to alienate him from his family, According to Eley, he resorted to crude abuse which isolated him still further from those closest to him who could have provided him with empathy and support.[11] This scenario is typical for people who are living with PTSD.

Whether or not George was living with depression and whether or not Jack had PTSD both of the men stranded on Howick Island certainly had their emotional problems. Given their predicament, their similarities and also their differences made conflict inevitable.

36

What Happened Next?

The story of the stranding on Howick Island cannot be left at the point of Jack's rescue. There is not much recorded about George's life after Howick – he went back to prospecting and mining and he was dead seven years later. Jack Idriess lived for another fifty-eight years to the grand age of eighty-nine and during that long life he established himself as one of Australia's most famous authors.

In February 1921 when Jack arrived back in Cooktown, George continued living on Howick Island. Following Jack's report to the Cooktown Police, Constable Brown travelled to Howick some time around 18 February 1921 and returned some time prior to 28 February 1921.[1] When Constable Brown returned he reported that George had food "*on which to subsist*".

At that time George still refused to leave the island but he did leave within a few months because by October that same year the Cooktown Mining Warden reported that George and five other men were prospecting and trenching several different claims. The Warden said George's gold mining field could not be considered payable because all the stone had to be transported to Cooktown for crushing.[2]

By May of the following year (1922) the Warden reported that "*Tritton and party*" had sunk a gold mining shaft to over seven metres but the shaft had filled with water.[3] George did not seem to be having much luck.

He made the news again in 1925 but still seemed to be missing the real gold.

The Assistant Mining Registrar at Coen reported gold finds (for everyone, it seemed, except George) at the Batavia diggings. He wrote, "*George Tritton has also bottomed but does not appear to have struck the real gutter as yet but gets good prospects occasionally. He got a good specimen, with about an ounce of gold in it a few days ago.*"[4]

In 1925 George was registered on the electoral roll as a Miner in Batavia River, Coen[5]

In 1927 George was at Spion Kop, Yarraden[6] but he would have been ill. Almost certainly that would have been his last mining venture and it is doubtful that he would have been able to do much work. George had developed neuritis which is today called neuropathy or peripheral neuropathy.

Neuritis is an extremely painful swelling and inflammation of specific nerves or groups of nerves. The result is weakness of nearby muscles, loss of reflexes and changes of sensation – particularly the sensation of heat or cold. One of the main symptoms is burning or stabbing pain but in severe cases there can be paralysis of nearby muscles.

Neuritis can be caused (among other factors) by a poor diet, an unhealthy lifestyle, nutritional deficiencies and alcoholism. These factors can upset the acid/alkaline balance of the blood. Perhaps George's lifestyle finally caught up with him but the serious gastro-intestinal condition with which he had been living since the First World War could also have been a contributing factor.

By 1928 George was in Cooktown General Hospital. He seems to have been there for about ten days before he died on 6 March 1928. He was fifty-four years of age. He was buried next day in the Cooktown Cemetery.

In George's death certificate, the use of numbered diagnoses: "(1) Neuritis" then "(2) Debility" then "(3) Cardiac failure" was obviously intended to indicate a sequence of events as the cause of death. George's heart failed but the reference to debility means that at only fifty-four years of age his body must have been feeble and exhausted. George was in a bad way. His overall weakness was certainly related to the primary diagnosis of

neuritis. The condition would have been creeping up on him. It would have been gradually worsening over the last few years of his life.

There is no more record of the detail of George's life after the Howick adventure but a great deal is known about the next fifty-eight years of Jack's life. In 1921 when he was rescued from Howick Island, the adventurous part of Jack's life was not quite over but his stranding on Howick Island did mark the beginning of a change in his career from adventurer to author.

As summarised in earlier Chapters, prior to the Howick adventure Jack had been wandering New South Wales and Queensland prospecting for opals, tin and gold. He had tried his hand at many rural jobs and he had fought through Gallipoli and the Middle East in the First World War. After he was rescued from Howick Island Jack continued his life of prospecting and he wandered north to the Torres Strait. However, he had already begun writing *Madman's Island* based on his experiences on Howick. Within a few more years his life of free adventure was virtually over.

It is quite reasonable to say that Jack's career as an author really began in the bar of the West Coast Hotel in Cooktown. Jack already had long experience as a diarist and he was a regular contributor to newspapers and magazines but his career as a published author began when he met George in the hotel bar and agreed to go prospecting on Howick Island.

It is not at all clear what Jack was doing between the time he arrived back in Cooktown in February 1921 and 1926 when he embarked on the *Somerset* to the tip of Cape York on a well-documented but frustrating prospecting mission.[7] Despite having access to Jack's diaries, Beverley Eley did not provide much information about Jack's experiences over this five-year period. This could have been the outcome of publication constraints.[8]

Eley said Jack was not at all well when he returned from Howick. She implied a psychological problem, "*...his nerves were stretched to breaking point.*"[9] Eley said Jack went "for a short time" to recuperate at the home of his sister Dycie (Ildyce) in Grafton, (New South Wales). Eley mentions several articles Jack published but says little about where he was living or what he was doing. The only chronological information Eley provides is that Jack had a Paddington (New South Wales) address in 1925, he was in Grafton later that year and he moved to Brisbane in 1926.[10]

However, in June 1921 less than four months after Jack was rescued it was reported that, "*J. Idriess and party are after gold and were last seen somewhere beyond the Starkie* (sic). *Jack will certainly strike it rich some day, or keep on trying.*"[11]

In 1923 Jack was in Brisbane. Around 27 March Jack gave a pint of blood to a seriously ill ex-serviceman in the Mater Misericordiae Hospital in Brisbane. On 29 March the Sydney Morning Herald and several other newspapers reported,

> "*SAVE A COMRADE. BRISBANE. Wednesday.*
>
> *To save a comrade's life, Private I. F. Idriess, late of the 5th Light Horse, has offered to give a pint of his blood. Several days ago the secretary of the Returned Soldiers' League (Major Dibden) received a request from Mrs. Pearce, wife of a Digger who had served with the Fourth Pioneers, stating that her husband was in a serious condition at the Mater Misericordiae Hospital, and that only the transfusion of a pint of blood would be likely to save his life. When Major Dibden made the news known it was not long before there were two volunteers. One man, who is in constant work, stated that he would forego his Easter holidays and make an attempt to save his comrade's life. Meantime, however, it was found that Idriess had reported direct to the hospital and had had a blood test taken. He states, even though he had never met him before, he will "give it a go" to save a Digger's life. Idriess was three times wounded at the war – when near Jerusalem, at Shrapnel Gully and at Lone Pine.*"[12]

It might also be possible to locate Jack by his press and magazine articles but this is not really reliable because Jack might (obviously) have been writing about one time and place when he was living at a different location. Secondly, many of his articles are fiction written simply to entertain. For example, in 1923 there was widespread newspaper speculation about a Marsupial Tiger killing stock on the Cape York Peninsula.[13] Jack's experience with this beast

was quoted in several newspapers but unfortunately this does not help to place him on the Cape because Jack's story was pure fiction. It had been published a year earlier in the Bulletin[14] and remarkably, in 1922 D.H. Lawrence blatantly plagiarised this story for his book *Kangaroo*.[15]

Having expressed this caution, in the absence of any other indication of where Jack was after he returned from Howick some of his Sydney *Bulletin* articles might give us a clue. For example, in an article published on 1 December 1921 Jack started with the words, "*My mate and I are camped in a lonely spot up near Cape York…*" Later that month another article began, "*There's a great argument raging on this field (Ebagoolah, NQ)…*".[16]

So drawing these pieces of (admittedly shaky) evidence together, except for a spell in Grafton early in 1921, Jack appears to have been in North Queensland between June 1921 through December to May/June of the following year (1922) and perhaps into 1923. Then he appears to have spent some time "down south". Around March 1923 he was in Brisbane and in April 1923 he had a draft of *Madman's Island* to submit to Angus and Robertson.[17] Jack might have been back on the Cape York Peninsula during 1924 – he appears to have been around the Normanby River at the end of that year.[18] Some time the next year (1925) he went to his Paddington address in Sydney but later that year he went back to Grafton. He moved to Brisbane in 1926.[19]

Jack's movements during the 1920s are not at all clear but it is possible that he was still devoting some of his time prospecting on the Peninsula but increasing his time around Sydney. However, he had more adventures ahead before he settled finally in Sydney and became an author.

In August 1926 Jack left on an ill-fated prospecting trip to the tip of the Cape York Peninsula. Reluctantly he travelled on the *Somerset*, a boat that was plagued with problems. As the mechanical problems increased so too did the irritability of the skipper. Glad it was over, Jack headed for Thursday Island where he found casual employment as a wharf labourer.[20]

By the time George died Jack had been working on Thursday Island and he had travelled through the islands of the Torres Strait. He had gathered the material for *Drums of Mer* (eventually published in 1933). Obviously using Jack's notes, Eley wrote that at the time George died Jack was travelling along

the Fly River in New Guinea. However, this information is sketchy and Eley also wrote that by the end of 1928 Jack was back in Sydney (albeit with a collection of New Guinea souvenirs).[21]

Soon after, in 1929, Jack thought he was going to die. He was diagnosed with cancer.[22] His belief that he was going to die (and nagging from his sister Ildyce) prompted him to retrieve his war diaries and begin the task of condensing them for *The Desert Column*. After finishing *The Desert Column* in late 1929 Jack returned to the Cooktown area to prospect the Laura River. Eley said it was here that he finally decided to live in Sydney and in Eley's opinion, Jack had turned his back on the bush so that he could get medical treatment.[23] He arrived in Sydney in February 1930.

Jack did not get married but he did have his own family. Around 1931 or 1932[24] when Jack was in his early 40s (and his stellar rise in public popularity had begun) he began a love affair with a married woman, Eta Gibson. Eta already had a child but for some reason Jack would never have anything to do with him.[25] Eta bore two other children (Judy and Wendy) from what became (according to Eley) a long-term "love-hate" *de facto* marriage.[26] Judy was born in 1932 followed by Wendy. So, albeit turbulent and stressful for all, Jack did finish up with a family of his own.

It was around 1930 that Jack finally turned his attention to full-time authorship but the writing and publication of his first book was not easy. Soon after he was rescued, Jack began writing *Madman's Island.* Eley said he began writing it at the home of his sister Ildyce in Grafton (New South Wales) and by 1923 he had completed a manuscript of his first book.[27]

As with so many episodes in Jack's life, there are contradictions and inconsistencies in the story of the book's publication. There were three players in the story – Jack, George Robertson of Sydney's Angus and Robertson (as distinct from the George Robertson of Melbourne's Robertson and Mullins) and Alec Chisholm. Chisholm was then news editor of the Sydney *Daily Telegraph* and a keen ornithologist. He had come to know Jack through Jack's magazine articles on the birds of North Queensland.

In the Author's Note to the second *Madman's Island*, Jack would have his readers believe that his first book was published merely by chance. Jack said he responded to Chisholm's suggestion that he should write a book by laughing derisively and saying: "*Not in my line. The nearest I have ever got to that is an old diary I kept on an island.*" This was disingenuous – not only was he a dedicated diarist but he was an experienced and frequently published writer of newspaper and magazine articles – Jack's first book was certainly not the product of an inexperienced writer.

He had written innumerable articles that were published in a variety of magazines and newspapers. The Sydney *Bulletin* alone published over 150 of his articles between 1911 and 1921 when he started writing *Madman's Island*.[28] Nevertheless, the idea pitched by Jack is that *Madman's Island* was "discovered" without him doing much at all. This is obviously not so.

One version of the book's publication was that while on a visit to Sydney, Jack was talking to Chisholm about birds when he mentioned his experience on Howick. He happened to have the manuscript of *Madman's Island* with him. According to this version, Chisholm was so excited by the story that he immediately took Jack across the road and introduced him to George Robertson. This version has Chisholm believing Jack had written the Great Australian Novel.[29]

Another version is set out in the Author's Note to the second *Madman's Island*. In this version the manuscript does not exist when Jack meets Robertson. Jack had with him only a "*rough diary written on an old sea log*". Discounting both those versions, a more likely scenario of the way that *Madman's Island* was published is the experience of most aspiring first-time authors.

Jack seems to have had to battle to get his work read. It certainly did not burst on to the Australian literary scene as the Great Australian Novel. Robertson's correspondence dated 11 April 1923 shows Chisholm simply enclosed the manuscript in a letter asking for Robertson's assistance on another matter. Robertson sent the manuscript back – unread.[30] Although Jack asked his friend Chisholm to intercede on his behalf, Chisholm did the absolute minimum. Chisholm did not deliver the manuscript to Robertson. He did not write a specific letter. He did not mention Jack by name. Nor did he in any way add his recommendation. He simply enclosed the manuscript with another letter to which he added that the manuscript was written under rather extraordinary circumstances by a bushman friend in Queensland.

So it seems that by 1923, the first version of Jack's first book had been rejected. Over the next few years Beverley Eley wrote that Jack continued to write short pieces for the Sydney *Bulletin* and for *The Australian Women's Mirror*. He also wrote a longer piece "*Heroines all – Women of the Far North – Great Heart and Dreams*".[31] Then he rewrote the manuscript of *Madman's Island* to include a fictitious love interest.

In the Author's Note to the second *Madman's Island*, Jack said he was directed by Robertson: "*Start from where this opium came ashore... be rescued as you tell here, but from there on have adventures with the opium and bring in a love interest quickly. Any beautiful girl will do. Marry her and I will take the responsibility*". In that same Author's Note Jack went on to write: "*Now (years later) the publishers have prevailed upon me to rewrite Madman's Island strictly according to fact, eliminating the fictitious love interest.*"

Directly following that Author's Note there appears an extraordinary "Publisher's Note". This is a rather testy justification. It seems to rebuke Jack for his temerity in laying blame on the publisher for the first book's failure. Also, by implication, Robertson seems to be denying blame for the inclusion of the love interest. "*We asked Mr Idriess, in 1927, to write the story as a novel, because, at that time, there was practically no demand for Australian 'true' adventure stories. In fact, twelve years ago almost any Australian book was a publisher's gamble.*"

Although Robertson (Eley wrote) never explicitly denied the love interest was his idea, Eley seems to support him and not Jack. She said a romantic storyline first appears in Jack's diary written in the Log of the *Sea Foam*. She wrote: "*It would seem that Robertson might have simply indicated that Jack should make more of the book's existing romantic aspect, not add it.*"[32]

Whoever was responsible for the distracting love interest, Jack's first book was published around June 1927 when he was working on Thursday

Island. Jack was just short of his 38th birthday. He had been working intermittently on his first book for well over four years (given that a first draft existed in 1923) and perhaps six years, if he started work in 1921 soon after his rescue. Although that book failed, the fact that it was accepted for publication encouraged Jack to continue writing. Subsequently he had a long and distinguished literary career.

There are various opinions about the actual number of Jack's books. For instance, in his authoritative work, *An Idriess Bibliography*, Ross Burnett puts the number at 59. Other assessments include that of the then Premier of New South Wales. In his Foreword to Beverley Eley's biography, Mr Carr put the total at "more than forty-seven books" and Eley herself has fifty as the total.

Jack published fifty-three books in forty-two years. In the list set out below, both *Madman's Island* books are included because the second *Madman's Island*, was re-written not just reprinted. Each of the Australian Guerrilla series is included because although small, they are distinct books in their own right. Not included are *The Battle for the Inland*, *Living off the Land*, or *Gems from Idriess*. These are works to which Jack contributed or which used his work. Not included are special or limited editions like *Flynn of the Inland*, *Gold Dust and Ashes* or *The Silent Service*. The list does not include; (a) collected editions of Jack's books, (b) foreign editions, nor (c) the many revised and enlarged editions of Jack's original work.

1. Madman's Island (1) 1927
2. Prospecting for Gold (Mar) 1931
3. Lasseter's Last Ride (Sep) 1931
4. Flynn of the Inland (Mar) 1932
5. The Desert Column (Apr) 1932
6. Men of the Jungle (Sep) 1932
7. Gold Dust… Ashes (Mar) 1933
8. Drums of Mer (Sep) 1933
9. The Yellow Joss 1934
10. Man Tracks 1935
11. The Cattle King 1936
12. Forty Fathoms Deep 1937
13. Over the Range 1937

14. Madman's Island (2) 1938
15. Cyaniding for Gold 1939
16. Must Australia Fight 1939
17. The Great Trek 1940
18. Headhunters of the Coral Sea 1940
19. Lightning Ridge 1940
20. Fortunes in Minerals 1941
21. The Great Boomerang 1941
22. Nemarluk: King of the Wilds 1941
23. Shoot to Kill 1942
24. Sniping 1942
25. Guerrilla Tactics 1942
26. Trapping the Jap 1942
27. Lurking Death 1942
28. The Scout 1943
29. Onward Australia 1944
30. The Silent Service 1944
31. Horrie the Wog-dog 1945
32. In Crocodile Land 1946
33. Isles of Despair 1947
34. The Opium Smugglers 1948
35. Stone of Destiny 1948
36. One Wet Season 1949
37. Wild White Man of Badu 1950
38. Across the Nullabor 1951
39. Outlaws of the Leopolds 1952
40. The Red Chief 1953
41. The Nor'westers 1954
42. The Vanished People 1955
43. The Silver City 1956
44. Coral Sea Calling 1957
45. Back o'Cairns 1958
46. The Tin Scratchers 1959
47. The Wild North 1960
48. Tracks of Destiny 1961
49. My Mate Dick 1962
50. Our Living Stone Age 1963
51. Our Stone Age Mystery 1964

52. Opals and Sapphires 1967
53. Challenge of the North 1969

Having fifty-three books published is a serious achievement for any author but if a dedicated collector set out to buy every edition of every one of Jack's books the total would be 350 and still counting. More than three decades after his death, (at the time of writing this book) there are still some of Jack's books in print.

Even so, this record must be set in the context of the Great Depression. At the height of Jack's popularity during the 1930s, readers still bought his books by the thousand. At that time almost every Australian had bought or read an Idriess book. For example, in ten of the twelve months of 1932 there was an Idriess book published – three new books, ten reprints and one second impression. There is no doubting the number of Jack's readers.

More than three million Idriess books have been sold and his readership was immense. For example, in 1975 the Public Lending Right gave authors a fee when a book was borrowed. This showed more Idriess books were being borrowed than any Australian author.[33] In another example of his popularity, Jack had been dead for more than twenty years when the Community Research Unit, Frankston City (VIC) conducted a lending library survey, "Frankston's Favourite Book". This survey placed Jack's fifty-years-old book *The Red Chief* in the top one hundred of popular books along with books by Asimov, Tolkien, Tolstoy, Dumas, Le Carre, Courtenay, Cussler and Patrick White.[34]

Yet today Jack and his work are either ignored or denigrated: even by people who should know better. It is true that most of his books have fallen from public popularity – it is not unusual for an author to fall out of fashion with the reading public as new ideas emerge and cultural imperatives evolve. Jack was at his peak during the Great Depression and before the Second World War. These national experiences changed Australia and no doubt changed the way people considered his work. No-one could have a problem with this. Popularity wanes for many reasons.

Objection can be taken, however, to the fact that in the literary history of Australia, Jack's place is (more often than not) denied or overlooked.

Sadly, there are many examples of contemporary literary critics who have either ignored or denigrated Jack's work. Few have acknowledged his place in Australian literature.[35]

Jack's name should be included in every list of Australia's notable authors if for no other reason than his popularity with the book-buying public. Even including contemporary authors, Jack remains one of Australia's best-selling authors and he sold millions of books in one of the most difficult periods of Australia's literary history – the Great Depression of the 1930s, the Second World War and the post-war period. Further, Jack's work has enjoyed continuing popularity across generations. Most of his books have been reprinted and there is a strong market for his used books.

Needless to say, the number of books an author sells is not necessarily a measure of the quality of the writing but it could be argued that no literary work has any value at all except the value that <u>readers</u> bring to it. Jack's work has been criticised and it is true that he writes as though he and his reader were yarning by a camp fire. His style and his use of language offend purists but his readers, demonstrably, value his books. In marked contrast to his literary critics Jack's readers have voted in the most measurable way – they bought his books. Nowhere is this illustrated more convincingly than in the genesis of *Men of the Jungle*. H.M. Green (Fisher Librarian) edited the book and told Jack:

> "*Time after time you seem to be copying the style of the cheapest and worst sort of wild-west 'thriller'... My notes won't help this book much; I haven't marked half or a quarter of what I don't like, and you would probably only spoil what is still a most interesting book if you tried to rewrite it*".[36]

Men of the Jungle was published in September 1932 and sold 6000 copies in a few months.[37] It ran to twenty new editions and reprints. It was published in Sweden, Denmark and France and included in three sets of Jack's collected works. Twenty-three years after it was first published it was still in print.[38] Enough said!

Men of the Jungle was one of the wide range of Jack's books. While some of his later work can be categorised as travel writing (a term used by modern critics to denigrate Jack's work), his books on prospecting and mining are still today acknowledged as being practical and authoritative texts. Jack's first successful book was *Prospecting for Gold.* His books on war and military strategy are based on his personal experience and some continue to be published. Some of Jack's ideas on the development of Australia are still being debated half a century after publication. Then he also wrote five books that were more-or-less biographical. Two of these are biographies of Aboriginal people.

Regardless of modern opinion about Jack and his work it must be said that most well-published and/or highly regarded Australian authors would regard the title "author" as an adequate summary of their life's worth. Jack would not want his life to be summed up so narrowly. He was "out there" and did things. Jack always wanted to be known as a man who wrote from practical experience. He would want his life to be measured (in some order or other) by his experiences as a soldier, a prospector/miner, explorer and only then as an author. More than likely, he would be happy to be labelled a man of the bush – a bushie.

Sometime in Australia's future it is absolutely certain that Jack's books will be again consulted as a record of an Australia long gone. Whatever the criticisms of his work, these are books written from diaries in which Jack recorded his observations of remote Australia and its people. At the level of eye-witness chronicles his books must be regarded as an invaluable part of Australia's history. No matter the influence of television, the internet and forms of communication still being devised, Jack's books will still be there awaiting re-examination. One year before Jack was born, Thomas Carlyle wrote of the value of the written word in history.[39]

Carlyle said the art of writing is the most miraculous of all things man has devised. He named as "*precious*" and "*great*" the mighty fleets and (presaging modern cities) "*vast many-engined cities*" but then he asked the question, "*What do they become*?" He compared the ruins of ancient Greece ("*dumb mournful wrecks*") to books in which ancient Greece is still literally alive and can be called up again and again.

This will be true of Jack's work. The Australia of which he wrote has changed and (perhaps because of this change) his books are no longer

fashionable. Certainly today some of his language is politically incorrect. But through his work (and the work of his contemporaries), the Australia of the first quarter of the Twentieth Century is still alive. Paraphrasing Carlyle, in Jack's books lies the soul of Australia as it then was. In Carlyle's words, "(such books) *are the articulate audible voice of the Past, when the body and material substance of it has altogether vanished like a dream*".[40]

37

The Truth about Charlie

In determining the truth about "Charlie" the first decision is whether Jack himself was being truthful. Jack told his readers the second *Madman's Island* was written "*strictly according to fact*"[1] Amazingly, Cooktown resident and historian, Jim McJannett, actually spoke with Jack about *Madman's Island.* During that conversation (of course in a Sydney pub) Jack said, "*The first book was all but fiction and the second book was ninety-nine percent fact*".[2] Neither of those statements were true. Even leaving aside embellishments and exaggerations inserted into the narrative for dramatic effect, Jack's claim that the second *Madman's Island* was ninety-nine percent fact is demonstrably untrue. For example, Jack fabricated the story of the wreck of Tarquay's boat the *Sea Foam* and the way he discovered the *Sea Foam* Log.

Even so, the <u>basic story-line</u> seemed to be confirmed by a press report of Jack's interview with the police when he returned to Cooktown in February 1921. On 16 February 1921 the *Brisbane Courier* reported:

> "*Prospector's Hardships.*
>
> *The Commissioner of Police has received a message from Cairns stating that the Cooktown police had reported, that Ian (sic) Llewellyn Idriess, a returned soldier and a mining prospector, had reached there from No. 1 Howick Island, 70 miles north of Cooktown, where he and George Tritton (also a returned, soldier and prospector) had been prospecting since September last, with poor results. Their provisions, with the exception of flour, had been used up, and for two months they had been living on fish. The island was otherwise uninhabited. Idriess signalled a passing fishing*

lugger and got away; but he stated that Tritton refused to leave. Tritton was said to be suffering from the effects of a war wound to his stomach."[3]

Almost word-for-word this story was picked up over the following days by the Melbourne *Argus* and other newspapers. These reports confirm Jack's report to the Cooktown police. They establish "Charlie's" real name – George Tritton. They confirm the fact that Jack and George did go prospecting on Howick Island. The dates of their stranding are quoted by Jack as September to February. They establish that George remained on the island after Jack was rescued. Finally, the reports confirm Jack's story that George was suffering the effects of an injury to his abdomen. These brief press reports place beyond doubt that the stranding on Howick did occur but there is room to doubt some of the other assertions in Jack's account of the adventure.

Many of Jack's readers doubted the truth of the story that was the foundation of the two *Madman's Island* books. Even Beverley Eley seemed to accord Jack's story little credibility. In her 370 page book Eley, discussing Jack's life-threatening adventure on Howick Island devoted just two pages that seem to have been drawn from a quite sad, often incoherent ABC interview provided to Tim Bowden when Jack was over eighty years old.[4]

Another critic of both of Jack's books is Jim McJannett of Cooktown. For decades Jim has devoted much time and effort into researching the Cape York Peninsula and its people. He spoke personally with Jack and he knows people who knew George Tritton. Jim respects Jack's ability as a story-teller but says his two *Madman's Island* books are just that – good stories. Jim dismisses even the basic premise of the story – the idea that the two men went prospecting for just one month then were stranded for a further four months. Jim has an encyclopaedic knowledge of people, localities and mining in North Queensland and he refers to an article dated 18 February 1921 in the *Cairns Post*. Several points in this article challenge Jack's statements that the second *Madman's Island* is true (or at least ninety-nine percent true). If the *Cairns Post* article is accurate then not just the first book but both of the *Madman's Island* books can be dismissed as fictional. Several elements of the *Cairns Post* article strike at the heart of Jack's published version of the story. The *Cairns Post* article reads:

Cairns Post (Qld) Friday 18 February 1921. Alone on Howick. Case of Returned Soldier. A Strange Affair. News Received in Cairns.

Inspector Brosnan, of the Cairns Police District, has received word from Cooktown of a rather strange case in connection with two returned soldiers named George Tritton, and J. L. Idriess. It appears that in September last, the two men, who had been prospecting in the Cooktown district, left there for No. 1 Howick Island in the ketch Ella, when she was on her way to Stewart River with .supplies for the Coen.

On the evening of the 6th inst, one of the men, Idriess, returned to Cooktown in a beche de mer boat, and made the rather sensational statement that there was a possibility of Tritton perishing for the want of supplies, or perhaps doing away with himself as, when Idriess left the island, his mate was certainly acting in a strange manner. Tritton at that time had just one bag of flour left, and was engaged in spearing fish for food.

In the course of an interview with the Cooktown police, Idriess said he was a prospector. "Last September", he continued, "Tritton and myself left Cooktown on a prospecting tour of No. 1 Howick Island. We left Cooktown in the cutter Ella, and arrived safely at the island.

We had taken with us what we thought would be provisions to last us for six months, but those supplies only lasted a bare four months, except for the flour, which we managed to make last all the time. During the first four months we got about 4 cwts. of tin and 4 cwts. of wolfram. Then as our supplies were fast diminishing, except the flour, we had to knock off work and spend our time fishing for tucker, so that for the last three months we have lived on nothing but flour and fish and crabs which we speared. About five days ago a Japanese beche de mer boat, in answer to my signals landed a dinghey, and took me off the island. We had half a bag of flour left at that time. Tritton, who remained on the island, has also about 25 chickens which he has carefully reared. He has also planted a vegetable garden. He says the fowls later on will keep him in meat, the garden in vegetables, and with what fish he can spear he will live and die on the island. He is

suffering very much from a wound in the stomach which he received in the war. In my opinion he is strange in his head. On two separate occasions he was partially out of his mind for a week on end. Then he pulled himself together again, and became quite normal.

He has a great deal of pain in his stomach, and in my opinion this trouble has affected his head. There is no one else on the island. During the whole time, we were there the captain of a cutter landed once, and a party from the survey ship Fantome landed for a few days. There are no blacks on the island, and no chance of a boat calling unless one is specially sent out.

When I left the island my mate gave me no authority to order any stores from Cooktown. During the last three months I tried to persuade Tritton to leave the island whenever a boat would call in answer to my signals, but he would not hear of leaving. He said he might just as well die on the island as in the Cooktown Hospital. He added that civilisation had never done anything for him, and he was sick of the rotten ways of his fellow men."

Speaking to a representative of the "Post" on Thursday, Inspector Brosnan said instructions had been issued for a boat to leave Cooktown forthwith and proceed to Howick Island to bring off Tritton.[5]

The main contradictory point in the interview reported by the *Cairns Post* was that George and Jack had taken supplies for six months. In the article Jack was quoted as saying these supplies had run out after four months. Given that the stranding was for a maximum of five months (September 1920 through January 1921), this piece of information means the men had sufficient supplies up to the end of December 1920 and even into January 1921. Jack was taken off the island a few weeks later. This is in direct contradiction to the hardship described in the *Madman's Island* books. Indeed, if the *Cairns Post* interview is correct, the men were not "stranded" at all; they always intended to stay for six months. In both *Madman's Island* books Jack said their intention was to stay for just one month and they had supplies for just this period of time.

The second part of the *Cairns Post* conundrum is the mention in the article that George still had twenty-five chickens on the island when Jack was

taken back to Cooktown. This information also strikes at the heart of the basic story. To some extent it might be confirmed by the finding of Constable Brown who went to Howick to take George off but found that he had food enough to subsist.[6] The existence of chickens on the island as another food source demolishes another theme of the *Madman's Island* books. In neither of the *Madman's Island* books was there mention of "Charlie" keeping chickens on the island but in both books the men did discuss keeping chickens. These discussions were in the context of "Charlie" talking about living permanently on the island and importing chickens and goats from Cooktown.

The third part of the conundrum is that Jack is quoted as saying an unnamed skipper in a boat called the *Ella* dropped them at Howick whereas Beverley Eley (presumably from Jack's notes) said it was Captain Moynahan in the *Spray* who delivered them to the island.[7]

So if (according to the *Cairns Post* interview) the two men had supplies for most of their time on Howick and they always intended to stay for six months, (a) they were not stranded and (b) they had to live off the resources of Howick for a much shorter time than Jack claimed in his books. If the *Cairns Post* report is accepted as factual all the drama of the story set out in Jack's books is destroyed. Either the *Cairns Post* report was true or the second *Madman's Island* book was written (as Jack claimed) strictly according to fact. There is no way of matching one with the other.

Jim McJannett's preferred scenario is that George took enough supplies for six months but just for himself. Jim's view is that Jack intended to stay for one month to help George set up a mining operation (as distinct from a prospecting trip). Jim said the two men would not have gone to the island for just one month. Then when the boat did not return on time (Jim believes) George had to share his supplies with Jack and they ran out after four months. This scenario sounds feasible but it completely undercuts the basic premise of the story of *Madman's Island* – that after the scheduled month had passed the two men were stranded for another four months and had to live off the island's resources. To accept this scenario is to hold that the whole story of the Howick adventure is fiction. Either the basic story in the books is factual or the *Cairns Post* article is accurate. They can't both be correct.

It is more likely the *Cairns Post* article is, for some reason, false.

One problem for the *Cairns Post* article is the chain of possession of the facts. Jack gave a statement to the police. The police relayed information to "a representative" of the paper. The representative's report was sent from Cooktown to Cairns and no doubt, subjected to the tender mercies of an editor. Somewhere during this chain of possession the paper decided to call the story "a rather strange case" and to refer to Jack's statement as "rather sensational".

Another curious fact about the *Cairns Post* article is that it seems to be quoting from Jack's statement to the police. This might be so but police statements are rarely taken in the first person. It is more likely the police statement would have been phrased in drier third-person officialese.

The *Cairns Post* article quotes Jack but later he wrote two books that contradict it. That Jack embellished the story for literary effect does not destroy the basic facts of the Howick adventure but the question remains: why would Jack tell one story to the police (or the *Cairns Post* representative) and another in his books? It could not have been a memory lapse. Although the first book was not published until six years later, Jack began writing it almost immediately after his rescue.

One measure of support for accepting the version set out in Jack's books is in Jack's commitment to the basic story. He returned to the story nearly ten years after the first book failed. Many successful authors have had a first time failure and this was Jack's experience. The first *Madman's Island* failed. Only 300 copies were sold at sixpence per copy.[8]

However, his second book, *Prospecting for Gold*, was an immediate success. Two thousand copies were sold in the first ten minutes after it was released for sale.[9] Then, before Jack returned to the story of *Madman's Island*, he had published another twelve books that were also selling in their thousands. He had another mining book in line for publication when the second *Madman's Island* was released. War was again in the air and he was turning his attention to the defence of Australia. In the next few years he would publish books based on his experience as a soldier.

Returning to a failed book must surely have seemed a huge risk to a now established reputation. The reason Jack returned at all to *Madman's Island* seems obvious. Jack's stranding on Howick with George was probably one of the most memorable periods in his adventurous life.

There are modern-day indicators for the truth of the basic story. Peter and Helen Rutherford of Cooktown have examined Jack's second *Madman's Island* by comparing the book directly with the features of Howick Island. They have walked the island with the 1938 book in hand, comparing Jack's words with the features of the island today. While they acknowledge inconsistencies in Jack's story, they have no doubt about the elemental truth of it. Peter said, "*It's all true. Everything he described on the island is still there*".[10]

It is impossible to reconcile the two contradictory positions and it is impossible to explain away the conundrum presented by the *Cairns Post* article. Nevertheless, Jack was an emotional mess when he returned to Cooktown. His account might not have been as coherent as it could have been and parts of his explanation might have been open to misinterpretation. There could have been misunderstandings between Jack's spoken words and the report as transcribed and published in the newspaper. There might even have been some self-justification in Jack's description of his arrival back in Cooktown when his mate was left alone on an island.

So was the story of *Madman's Island* really "true"? After finding Charlie's real name, comparing Jack's two *Madman's Island* books, talking to Jim McJannett about George's life and talking to Peter and Helen Rutherford about Howick Island, the essential elements of the story can be regarded as "factual". Nevertheless, it is also clear that Jack did elaborate his story for dramatic effect. The second *Madman's Island* book is mostly factual – not ninety-nine percent – <u>mostly</u> factual. Returning to the words of Mark Twain in his book *The Adventures of Huckleberry Finn*, there were things which (Jack) stretched, but mainly he told the truth.[11]

Nevertheless, Jack was incorrect when he declared "Charlie" (George Tritton) to be mad. George was an unusual man. His personality was developed through his childhood living on the harsh Devon coast. This childhood provided him with a tough, resilient independence. It prepared

him for a life of struggle and hardship in the jungles of the Cape York Peninsula. His capacity for hard physical work and his attitude toward sharing camp responsibilities differed from Jack's admitted lackadaisical attitudes. In retrospect there was never going to be any way that these two men could have lived together in enforced isolation without conflict. But it is not likely that George had a psychiatric disorder – he was not "mad".

It is not possible to explain (or excuse) George's acts of aggression toward Jack. He certainly did fire the rifle at Jack but he had just taken a beating from a younger man who had some experience as a boxer. George did start that fight (according to Jack) and he apparently did start the earlier fist fight. The point is, however, that these acts of aggression were provoked. They were not sudden maniacal rages. George had been steadily more frustrated with Jack's slovenly camp habits – beginning even from day one. George's acts of aggression also have to be set against Jack's behaviour. It was Jack who burned George's tent and (perhaps) everything in it. It was Jack who deliberately tried to kill George with a rock. And Jack did not help George when he thought George was in danger in the mud-pool.

It is also easy to forget the context of the conflict. The reader sitting comfortably reading a copy of *Madman's Island* finds it hard to imagine a context in which two men are alone on a hot, nearly dry, island only the size of about twenty suburban house blocks. For five months.

Given Jack's behaviour and given the frequent provocation, George's acts of aggression cannot be regarded as symptoms of a mental illness. Nor can his communication style (that Jack frequently described as growling) be regarded as a symptom. George was too introspective and taciturn to suit even the quiet Jack. This got on Jack's nerves it does not make George "mad". Earlier in this book Jack's experience with the Jungle Man was described[12]. This showed that the Jungle Man also got on Jack's nerves. None of this seems to warrant the label of "madness".

Over six years before George died he would have read the *Cairns Post* article with its derogatory references to his mental state. He would have been angry at that time. Then Jack's book appeared in Cooktown. It must be more than likely that George read Jack's book before he died. The book would have been widely read and discussed in Cooktown for about seven months before

George died. George Tritton (even when nominated as "Charlie") would have become known as the "madman" of *Madman's Island.* The melancholy conclusion of this book is that George was dying when he was wrongly branded as a madman and he took this unfair label to his death.

This sad thought is the truth about Charlie.

Vale – Ion Llewellyn (Jack) Idriess – 1889-1979

Vale – Thomas George Tritton ("Charlie") – 1874-1928

Endnotes

The Facts

1 The *Brisbane Courier* (Qld), Monday 16 February 1921, p 6. The *Brisbane Courier* (Qld), Monday 28 February 1921, p 6.
2 See Chapter Three – Howick Island.
3 The *Brisbane Courier* (Qld), Monday 16 February 1921, p 6. *Cairns Post* (Qld) Friday 18 February 1921
4 The *Brisbane Courier* (Qld), Monday 16 February 1921, p 6. *Cairns Post* (Qld) Friday 18 February 1921, p 4. The *Brisbane Courier* (Qld), Monday 28 February 1921, p 6. *Cairns Post* (Qld), 20 October 1921, p 4.
5 Idriess I, 1927, *Madman's Island,* Cornstalk, Sydney. Idriess I, 1938, *Madman's Island,* Angus and Robertson, Sydney.
6 Idriess I, 1938, *Madman's Island,* Angus and Robertson, Sydney, p VII..
7 *Cairns Post* (Qld) Friday 18 February 1921, p. 4. See Chapter Thirty-seven.
8 Clemens, S. L. (Mark Twain) (1885), *The Adventures of Huckleberry Finn*, Charles L Webster and Company, NY.
9 Bowden, T 1972? *http://www.abc.net.au/rn/radioeye/documents/idriess.pdf*
10 Eley B, 1995, *Ion Idriess*, Harper Collins, Sydney.
11 Dixon R, 1996, 'Biography in a Butterbox', *Coppertales* No. 3, November 1996, p 117. Corris P, 1995, 'The King of Derring-do', *Sydney Morning Herald*, 16 June 1995, p8. Stephens T, 1995, 'Idriess – Angel and Devil', *Sydney Morning Herald*, 4 August 1995, p 12. Ryan P, 1995, 'Alas, poor Jack…', *Sydney Morning Herald*, 4 August 1995, p 10.
12 Roderick C, 1947, *Twenty Australian Novelists*, Angus and Robertson, Sydney. Foster D, 1991, *Self Portraits*, National Library of Australia, Canberra. Hetherington J, 1963, *Forty-two Faces*, Angus and Robertson, London. Ruhen O, 1979, 'Frontiersman with a Pen', *Australian Book Review*, September 1979, p 30. Casey G, 1964, 'A Handshake for Ion Idriess', *Meanjin Quarterly*, April 1964, p 348. de Berg H, 1992, *Ion Idriess*, Audio interview, National Library of Australia. Dyson J, 1997, 'Cyclone Jack Idriess', *Readers Digest*, July 1977, p 99. Croft J, 1983, 'Idriess, Ion Llewellyn 1889–1979', *Australian Dictionary of Biography*, National Centre of Biography, Australian National University. De La Rue K, *http://delarue.net/idriess.htm.*
Wikipedia, *http://en.wikipedia.org/wiki/Ion_Idriess.*

Preface

1 Idriess, I 1927, *Madman's Island*, Cornstalk, Sydney, p 75.
2 Idriess, 1927, p 75.
3 Eley, B 1995, *Ion Idriess*, Harper Collins, Sydney, pp 78, 80, 96.
4 Idriess, 1927, pp 10-11. Idriess, I 1938, *Madman's Island*, Angus and

Robertson, Sydney, pp 29-30. Defence Service Records, 'Tritton, George', *National Archives of Australia.*

Chapter One - George's Early Life

1 ancestry.com.au, 'Thomas G. Tritton', *England Census 1881.*
2 ancestry.com.au, 'William F. Tritton', *England Census 1881.*
3 ancestry.com.au, 'Elizabeth Tritton', *England Census 1881.*
4 ancestry.com.au, 'William F. Tritton', *England Census 1881.*
5 genuki.org.uk. 'William Tritton'. http://www.genuki.org.uk/big/costguards.
6 genuki.org.uk. 'William Tritton'. http://www.genuki.org.uk/big/costguards.
7 le Cheminant R, 1999, *Hope Cove, The History of a Devonshire Fishing Village*, Self-published, London, p 1.
8 le Cheminant, 1999, pp 86-91.
9 le Cheminant, 1999, p 86.
10 le Cheminant, 1999, p 31.
11 le Cheminant, 1999, p 43.
12 le Cheminant, 1999, p 46.
13 le Cheminant, 1999, p 43
14 le Cheminant, 1999, pp 33-34.
15 le Cheminant, 1999, pp 36-37.
16 le Cheminant, 1999, p 55.
17 le Cheminant, 1999, pp 55-56.
18 le Cheminant, 1999, pp 12-13.
19 *www.angelfire.com/ga/BobSanders/Coastguards.html*
20 le Cheminant, 1999, p 13.
21 www.coastguardsofyesteryear.org/forum/viewthread.php?thread_id=562
22 le Cheminant, 1999, p 54.
23 genuki.org.uk. 'William Tritton'. http://www.genuki.org.uk/big/costguards.
24 pension100.co.uk, *www.pension100.co.uk/historyofpensions/timeline.htm.*
25 ancestry.com.au, 'William F. Tritton', *England Census 1881.*
26 Holthouse, H 1967, *River of Gold*, Angus and Robertson, Sydney, pp 186-187.
27 Pike, G 1981, *The Golden Days*, Pinevale Publications, Mareeba, p 89.
28 Pike, G 1979, *Queen of the North*, Self published, p 7.
29 *Index to Registers of Immigrant Ships' Arrivals*, Tritton, 1848-1912.
30 Pike, G 2001, *Unsung Heroes of the Queensland Wilderness*, CQU Press, p 90.
31 ancestry.com.au, 'William F. Tritton', *England BMD Index* 1837-1915.
32 genuki.org.uk. 'William Tritton'. http://www.genuki.org.uk/big/costguards.

Chapter Two - Jack's Early Life

1 Eley, B 1995, *Ion Idriess*, Harper Collins, Sydney, p 9.
2 ancestry.com.au, 'Walter O Idriess', *Australia Marriage Index*, 1788-1950.
3 ancestry.com.au, 'Walter O Idriess', *Australia Death Index*, 1787-1955.
4 New South Wales Registry, 1889, *Births, Deaths and Marriages*, 'Ion Windeyer'.
5 Eley 1995, pp 11,19.

6 http://www.airgale.com.au/windeyer/d7.htm
7 ancestry.com.au, 'Julia M Edmonds' *Australian Birth Index*, 1788-1922.
7 Eley 1995, p 11.
8 Eley 1995, p 12.
9 ancestry.com.au, 'Julia M Edmonds' *Australian Birth Index*, 1788-1922.
10 New South Wales Registry, 1889, *Births, Deaths and Marriages*, 'Ion Windeyer'.
11 de Berg, H 1974, *Ion Idriess*, National Library of Australia, Audio interview.
12 *Barrier Miner* Broken Hill, 7 September 1950, p 4
13 Idriess, I 1938, *Madman's Island*, Angus and Robertson, Sydney, p VII.
14 *The Argus*, Melbourne, 22 March 1933, p 13.
15 Idriess, I 1958, *Back o' Cairns*, Angus and Robertson, Sydney, Author's Note.
16 Bradly, J ed. 2008, *Gouger of the Bulletin*, Idriess Enterprises, Uralla, NSW.
17 Idriess, I 1963, *Our Living Stone Age*, Angus and Robertson, Sydney, p 194.
18 Idriess, I 1959, *The Tin Scratchers*, Angus and Robertson, Sydney, p 72.
19 Idriess, I 1940b, *Lightning Ridge*, Angus and Robertson, Sydney, pp 6-12. Idriess, I 1956, *The Silver City*, Angus and Robertson, Sydney, pp 2-28.
20 Idriess 1940b, p 9.
21 Idriess 1940b, pp 9-10.
22 Idriess 1956, pp 10-11.
23 Idriess 1956, pp 9-10.
24 Eley 1995, p 20.
25 Idriess 1940b, p 13.
26 Idriess 1940b, pp 14-15.
27 Idriess 1940b, p 15. *Barrier Miner*, 7 September 1950, p 4.
28 *Barrier Miner*, 18 January 1907, p 4.
29 *Barrier Miner*, 18 January 1908, p 4.
30 *Barrier Miner*, 26 May 1904, p 2.
31 *Barrier Miner*, 14 January 1908, p 2.
32 Idriess 1940b, p 38. Eley 1995, p 28.
33 Eley 1995, p 27.
34 Eley 1995, p 27.
35 Idriess 1940b, p 39.
36 Idriess 1940b, pp 39-41.
37 Idriess 1940b, p 36.
38 Idriess 1940b, pp 40-41.
39 Idriess 1940b, p 39.

Chapter Three - Howick Island

1 Lucas, A 1968, *Cruising the Coral Coast*, Horwitz, Cameray, p 14.
2 *Brisbane Courier*, 2 June 1892, p 4.
3 Chapter Twenty.
4 Lucas 1968, p 16.
5 Idriess, I 1938, Madman's Island, Angus and Robertson, Sydney, p 192.
6 Lucas 1968, p 16.

7 Idriess, I 1927, *Madman's Island*, Cornstalk, Sydney, p 5. Idriess 1938, p 10.
8 Collingridge, V 2002, *Captain Cook*, Random House, London.
9 Gill, J 1979, *Journal of the Royal Historical Society of Queensland,* Lieut. Charles Jeffreys, RN – The Last Buccaneer? Vol 10 issue 4, p 100.
10 Currey, C 1963, *The Transportation, Escape and Pardoning of Mary Bryant*, Palmer, Sydney.
11 Horsburgh, J 1836, *Vol II, The India Directory*, W. H. Allen & Co. London, pp 791-792
12 Gill 1979, pp 98-122.
13 King, P 1827, *Narrative of a survey of the intertropical and western coasts of Australia, Volume 2*, John Murray, London, p 392.
14 Gill 1979, p 105.
15 Gill 1979, p 102.
16 Gill 1979, p 103.
17 Gill 1979, p 103.
18 Gill 1979, p 104.
19 Gill 1979, p 105.
20 Gill 1979, p 111.
21 Gill 1979, p 112.
22 Gill 1979, p 112.
23 Gill 1979, pp 115-116.
24. Gill 1979, pp 113-114, 115
25 Gill 1979, p 117.
26 Gill 1979, p 118.
27 Gill 1979, p 119.

Chapter Four - George and Jack

1 Idriess, I 1938, *Madman's Island,* Angus and Robertson, Sydney, pp 4, 192.
2 Idriess 1938, pp 2-3.
3 Idriess 1938, pp 4-5.
4 Idriess 1938, pp 2-3.
5 Defence Service Records, *National Archives of Australia*, Idriess, Ion Llewellyn.
6 Eley, B 1995, *Ion Idriess*, Harper Collins, Sydney, p 73.
7 Eley 1995, p 294.
8 Defence Service Records, *National Archives of Australia*, Idriess, Ion Llewellyn.
9 Eley, 1995, p 294.
10 The Courier Mail, Brisbane, 9 August 1950.
11 Defence Service Records, *National Archives of Australia*, Tritton, George.
12 Defence Service Records, *National Archives of Australia*, Tritton, George.
13 Idriess 1938, p 2.
14 Idriess, I 1927, *Madman's Island*, Cornstalk, Sydney, pp 74-75.
15 Idriess 1927, p 90.
16 Idriess, 1938, p 43.

17 Idriess, 1938, p 252.
18 Idriess, 1938, p 4
19 Idriess 1938, p 12.
20 Idriess 1938, p 22.
21 Idriess 1938, pp 4, 16, 20, 219.
22 Idriess 1938, pp 15-16.
23 Idriess 1938, p 104.
24 Idriess 1938, p 192.
25 Australian Electoral Rolls, 1901-1936, *George Tritton*, Ancestry.com.au.
26 *Cairns Post*, 20 October 1920, p 4. *Cairns Post*, 2 May 1922, p 8.
27 Idriess 1938, p 216.
28 Eley 1995, p 36.
29 Eley 1995, pp 80-81.
30 Idriess 1938, pp 53-54.
31 Idriess 1938, p 216.
32 Idriess 1938, p 54.

Chapter Five – Planning, June 1920

1 Eley, B 1995, *Ion Idriess*, Harper Collins, Sydney, pp 64, 109.
2 Idriess, I 1934, *The Yellow Joss*, Angus and Robertson, Sydney, p V.
3 Idriess, I 1962, *My Mate Dick*, Angus and Robertson, Sydney, pp 30, 37, 171.
4 Idriess, I 1959, *The Tin Scratchers*, Angus and Robertson, Sydney, e.g. p 38.
5 Defence Service Records, *National Archives of Australia*, Welsh, Richard Albert.
6 Idriess, I 1927, *Madman's Island*, Cornstalk, Sydney, p 2. Idriess, I 1938, *Madman's Island*, Angus and Robertson, Sydney pp 2, 3.
7 Eley 1995, p 100.
8 Idriess 1938, frontispiece.
9 Encyclopaedia of Australian Shipwrecks, *Queensland Shipwrecks, Sea Foam*, http://oceans1.customer.netspace.net.au/qld-wrecks.html.
10 Eley 1995, p 100.
11 *Cairns Post* Qld, 1 November 1918.
12 Idriess 1927, p 2.
13 Idriess 1938, p 3.
14 Queensland Registry, 1899, Births Deaths and Marriages, - *Oosop*, 1098, 2591.
15 *Brisbane Courier*, 9 June 1892, p 5.
16 Idriess 1938, p 125.
17 Idriess 1938, p 125.
18 Idriess, I 1958, *Back o' Cairns*, Angus and Robertson, Sydney, pp 97, 142-3.
19 Eley 1995, p 107.
20 Eley 1995, p 100.
21 Idriess 1927, pp 1,3. Idriess 1938, p 3, Ch XV.
22 Idriess, I 1963, *Our Living Stone Age*, Angus and Robertson, Sydney, p 59.

Chapter Six – "Near-in" Prospecting, 1919-1920

1 Personal communication with Jim McJannett. See Acknowledgements.
2 Idriess, I 1938, *Madman's Island*, Angus and Robertson, Sydney, pp 4, 16, 20, 87, 219,
3 Idriess, I 1958, *Back o' Cairns*, Angus and Robertson, Sydney, p 120.
4 Idriess 1938, pp 9, 10, 18. Idriess 1958, p 145 159.
5 Idriess 1958, p 142.
6 Idriess 1958, p 143.
7 Idriess 1958, pp 151, 153, 184, 185, 187, 189, 235, 241, 297, 262, 263.
8 Idriess 1958, p 199
9 Idriess 1958, pp 245, 249, 251.
10 Idriess 1958, pp 129, 174, 232, 236, 239, 273.
11 Idriess 1958, p 232.
12 Idriess 1962, pp 30, 37, 171.
13 Idriess 1962, p 8
14 Idriess 1962, pp 66/67.
15 Idriess 1962, p 223.

Chapter Seven – Landing, Septembre 1920

1 Chapter Twenty
2 Idriess, I 1938, *Madman's Island*, Angus and Robertson, Sydney, p 4.
3 Idriess, I 1927, *Madman's Island*, Cornstalk, Sydney, p 3. Idriess, I 1938, *Madman's Island*, Angus and Robertson, Sydney, p 5.
4 Idriess 1927, p 5. Idriess 1938, p 8.
3 Idriess 1927, p 5. Idriess 1938, p 10.
5 Chapter Three.
5 Idriess 1927, p 5.
6 Idriess 1938, p 9.
7 Idriess 1927, p 4. Idriess 1938, pp 7,9.
8 Idriess 1938, p 10.
9 Idriess 1927, p 4. Idriess 1938, p 9.
10 Idriess 1927, p 5. Idriess 1938, p 11.
11 Idriess, 1938, pp 12-17.
12 http://en.wikipedia.org/wiki/Mahina_Cyclone_of_1899.
13 Idriess 1938, p 13
14 Holthouse, H 1971, *Cyclone*, Angus and Robertson, Sydney, p 15.
15 http://www.em.gov.au/Documents/How_high_was_the_storm_surge_from_Tropical_Cyclone_Mahina.pdf
16 Holthouse 1971, pp 6-7.
17 http://en.wikipedia.org/wiki/Cyclone_Tracy.
18 http://en.wikipedia.org/wiki/Mahina_Cyclone_of_1899.
19 Idriess 1938, p 13.
20 Idriess 1938, p 15.
21 Idriess 1927, p 6.
22 Idriess 1938, pp 12-17.

Chapter Eight - George and Jack as Young Adults

1 Australian Electoral Roll 1903-1980, *Tritton Washington*, Ancestry.com.au.
2 Worker Brisbane, 16 September 1893.
3 Australian Electoral Roll 1903-1980, *Tritton Washington*, Ancestry.com.au.
4 Australian Electoral Roll 1903-1980, *Tritton William*, Ancestry.com.au.
5 Australian Electoral Roll 1903-1980, *Tritton George*, Ancestry.com.au.
6 Personal communication with Jim McJannett. See Acknowledgements.
7 Idriess, I 1940b, *Lightning Ridge*, Angus and Robertson, Sydney, p 48.
8 Idriess 1940b, p 49.
9 Idriess 1940b, pp 51-53.
10 Idriess 1940b, pp 54-55.
11 Idriess 1940b, pp 55-57.
12 Idriess 1940b, pp 58-59.
13 Idriess 1940b, p 63.
14 Idriess 1940b, p 65.
15 Idriess 1940b, pp 66-67.
16 Idriess 1940b, pp 68-71.
17 Idriess 1940b, p 71.
18 Idriess 1940b, p 73.
19 Idriess 1940b, pp 76-78.
20 Idriess 1940b, pp 81-82.

Chapter Nine - Prospecting, September 1920

1 Personal communication with Peter Rutherford. See Acknowledgements.
2 Idriess, I 1927, *Madman's Island*, Cornstalk, Sydney, p 7. Idriess, I 1938, *Madman's Island*, Angus and Robertson, pp 18-19.
3 Idriess 1927, p 7. Idriess 1938, p 19.
4 Idriess 1938, pp 16-17, 20.

Chapter 10 - Aboriginal People

1 Personal communication with Wilfred (Willie) Gordon. See Acknowledgements.
2 Personal communication with Wilfred (Willie) Gordon. See Acknowledgements.
3 http://en.wikipedia.org/wiki/Guugu_Yimithirr_language
4 Stanner, W 1968, *Boyer Lecture*, Australian Broadcasting Commission.
5 Gordon W, (with Bennett, J), *Guurbi. My Special Place*, Guurbi Tours, Cooktown.
6 Blackburn J, 1971, *Milirrpum v Nabalco Pty Ltd*, 58 1971 17 FLR 141, 267.
7 Chapter Fourteen.
8 Parkin, R 1997, *HM Bark Endeavour*, The Miegunyah Press, Melbourne, ch4.
9 Parkin 1997, p 328
10 Jack, RL, 1922, *Northmost Australia*, George Robertson & Co. Vol.1, pp 149-150.
11 Jack 1922, pp 237-242.

12 Coulthard-Clark, C 2001, *The Encyclopaedia of Australia's Battles*, Allen & Unwin, p 44.
13 Holthouse, H, 1967 River of Gold, Sydney: Angus & Robertson, pp 52-53, etc. Buchorn, R, *A Taste for Chinese?* The Skeptic Vol.14, No.1, p 30.
14 Idriess 1927, pp 3, 4, 17, 18, 79, 97, 98.
15 Idriess 1938, p 4, 16-17.
16 Idriess 1938, pp 16-17.
17 Idriess, 'Give 'Em A Go', *SALT*, vol. 5, no. 9, 1943, p 18.
18 Bradly, J ed. 2008, *Gouger of the Bulletin*, Idriess Enterprises, Uralla.
19 Miller, P 2002, Metamorphosis: Travel Narratives and Aboriginal/non-Aboriginal Relations in the 1930s. *Journal of Australian Studies*.
20 Shoemaker, A 1989, *Black Words, White Page*, UQ Press, StLucia, at p 54 quoted D.L.M. Jones's 1960, "*The Treatment of the Australian Aborigine in Australian Fiction*" and Healy, JJ, 1978, "*Literature and the Aborigine in Australia, 1770-1975*".

Chapter 11 - Exploring, September 1920

1 Idriess, I 1938, *Madman's Island*, Angus and Robertson, Sydney, p 22.
2 Idriess, I 1940b, *Lightning Ridge*, Angus and Robertson, Sydney, p 130. Idriess, I 1962, *My Mate Dick*, Angus and Robertson, Sydney, e.g. p 65.
3 Idriess 1938, p 21.
4 Idriess 1938, p 22.
5 Idriess 1938, pp 22-28.
6 Idriess 1938, p 22.
7 Idriess 1940b.

Chapter 12 - Lightning Ridge, 1909-1912

1 Australian Electoral Rolls, 1901-1936, *George Tritton*, Ancestry.com.au.
2 Idriess, I 1940b, *Lightning Ridge*, Angus and Robertson, Sydney, p 75.
3 Idriess 1940b, p 79.
4 Idriess 1940b, p 81.
5 *Barrier Miner* (Broken Hill, NSW) 26 May 1904, p 2.
6 *Sydney Mail*, 13 July 1910.
7 Idriess 1940b, p 108.
8 Idriess 1940b, p 111.
9 Idriess 1940b, p 112.
10 Idriess 1940b, p 128.
11 Idriess 1940b, pp 125-137.
12 Eley, B 1995, *Ion Idriess*, Harper Collins, Sydney, p 40.
13 Bradly, J ed. 2008, *Gouger of the Bulletin*, Idriess Enterprises, Uralla.
14 Idriess 1940b, pp 141-142.
15 Idriess 1940b, p 166.

Chapter 13 - Improvising, September 1920

1 Idriess, I 1927, *Madman's Island*, Cornstalk, Sydney, p 10. Idriess, I 1938, *Madman's Island*, Angus and Robertson, Sydney, p 28.
2 Idriess 1938, p 29.
3 Idriess, 1927, p 10. Idriess 1938, p 29. Defence Service Records, *National Archives of Australia*, Tritton, George.
4 *Cairns Post*, Friday 18 February 1921.
5 Defence Service Records.
6 Queensland Registry, 1928, Births Deaths and Marriages, *George Trittan* (sic), 287, 4229.
7 Defence Service Records.
8 Defence Service Records.
9 Idriess 1938, p 31.
10 Idriess 1927, p 11. Idriess 1938, p 31.
11 Idriess 1927, p 11. Idriess 1938 pp 32-33.
12 Idriess 1927, p 11. Idriess 1938 p 34.
13 Idriess 1938, pp 32, 48-49.
14 Idriess 1938, p 34.
15 Idriess 1927, pp 12-13.
16 Idriess 1927, p 13.
17 Idriess 1938, p 35-36.
18 Idriess, I. 1957, *Coral Sea Calling*, Angus and Robertson, Sydney, p 102.
19 Idriess 1938, p 36.

Chapter 14 - Cook's Luck

1 The Endeavour Journal of James Cook, 2008, National Library of Australia. Parkin, R. 1997, *HM Bark Endeavour*, The Miegunyah Press, Melbourne, pp 145-146.
2 http://atsiphj.com.au
3 Collingridge, V, 2002, *Captain Cook*, Random House, London.
4 Bowen J and Bowen M, 2002, *The Great Barrier Reef*, Cambridge University Press, UK, pp 13, 37-40.
5 Blainey, G 2008, *Sea of Dangers*, Penguin, Camberwell, pp 378-379.
6 Idriess, I 1948, *The Opium Smugglers*, Angus and Robertson, Sydney, p 24.
7 Idriess, I 1957, *Coral Sea Calling*, Angus and Robertson, Sydney, p 102.
8 Evans, D, 1969, (compiler), *An Account of a Voyage around the World with a Full Account of the Voyage of the Endeavour in the Year 1770 along the East Coast of Australia by Lieutenant James Cook*, Smith and Patterson, Brisbane, pp 212-215.
9 Besant, W 1890, *English Men of Action*, Macmillan & Co., London.
10 Parkin 1997, p 74.
11 Collingridge, V, 2002, *Captain Cook*, Random House, London.
12 Lucas, A. 1968, *Cruising the Coral Coast*, Horwitz, Cameray, p 278.
13 Lucas 1968, p 278.
14 Parkin 1997, p 311

15 Parkin 1997, p 312.
16 Lucas 1968, p 278.
17 Parkin 1997, pp 311-312.
18 Gill, J 1988, *The Missing Coast*, Queensland Museum, Brisbane, p 85.
19 Parkin 1997, p 312.
20 Parkin 1997, pp 320-321.
21 Parkin 1997, p 320.
22 Idriess, I 1962, *My Mate Dick*, Angus and Robertson, Sydney. Idriess, I 1958, *Back o'Cairns*, Angus and Robertson, Sydney.
23 Parkin 1997, p 368.
24 Parkin 1997, p 370.
25 Parkin 1997, pp 384-386
26 Lucas 1968, p 14.
27 Parkin 1997, pp 414-415.
28 Beaglehole, JC ed. 1962, *The Endeavour Journal of Joseph Banks*, vol II pp 105-6.
29 Parkin 1997, p 421.
30 Evans 1969, pp 212-215.
31 Idriess, I 1938, *Madman's Island*, Angus and Robertson, Sydney, p 36.

Chapter 15 – Visitors, September 1920

1 Idriess, I 1938, *Madman's Island*, Angus and Robertson, Sydney, p 43.
2 Idriess, I 1927, *Madman's Island*, Cornstalk, Sydney, pp 12-15.
3 Idriess 1938, p 34.
4 Idriess 1938, p 43.
5 Idriess 1938, pp 43, 48-49.
6 Personal communication with Peter Rutherford. See acknowledgements.
7 *Cairns Post*, Friday 18 February 1921.
8 http://en.wikipedia.org/wiki/HMS_Fantome_(1901).
9 Idriess 1927, p 83. Idriess 1938, p 202.
10 Idriess 1938, p 46.

Chapter 16 – Sand-flies, October 1920

1 Idriess, I 1938, *Madman's Island*, Angus and Robertson, Sydney, p 55.
2 Idriess 1938, pp 54-55, 57-59.
3 Idriess 1938, pp 49-50.
4 Idriess 1938, p 60.
5 *Sunday Times*, Perth, 3 July 1927, p 3.
6 Idriess 1938, pp 60-61.
7 Idriess 1938, p 63.
8 Idriess 1938, p 66.
9 Idriess 1938, p 65.
10 Idriess 1938, p 67.
11 Idriess 1938, p 66.
12 Idriess 1938, p 65.

13 Idriess 1938, p 67.
14 Idriess 1938, pp 68-69.
15 Idriess 1938, p 70.
16 Idriess 1938, pp 67.
17 Idriess 1938, p 68.
18 Idriess, I 1927, *Madman's Island*, Cornstalk, Sydney, p 18. Idriess 1938, p 68.
19 Idriess 1927, pp 18-23. Idriess 1938, pp 68-74, 75-79.
20 Idriess 1938, p 79.
21 Idriess 1927, pp 24-25. Idriess 1938, pp 81-82.
22 Idriess 1927, p 26. Idriess 1938, p 82.
23 Idriess 1938, pp 87-88.
24 Idriess 1938, p 89.
25 Idriess 1938, p 83.

Chapter 17 - King Tide, October 1920

1 Idriess, I 1927, *Madman's Island*, Cornstalk, Sydney, p 29.
2 http://www.msq.qld.gov.au/Tides/King-tides.aspx
3 Idriess 1927, p 29.
4 Idriess, I 1938, *Madman's Island*, Angus and Robertson, Sydney, p 99.
5 Idriess, I 1958, *Back o'Cairns*, Angus and Robertson, Sydney.

Chapter 18 - Cairns

1 See Chapter Four.
2 Australian Electoral Rolls, 1901-1936, *George Tritton*, Ancestry.com.au.
3 See Chapter Twelve.
4 Idriess 1958, p 6.
5 Eley, B 1995, *Ion Idriess*, Harper Collins, Sydney, p 43.
6 Pike, G. 1980 Pioneers' Country, p 139.
7 AAP, Brisbane, 24 Apr 2003.
8 Idriess, I 1958, *Back o' Cairns*, Angus and Robertson, Sydney, p 28.
9 Idriess 1958, pp 36-37.
10 http://www.herbertonvisitorcentre.com.au
11 *Cairns Post* QLD, 1 February 1939, p 11.
12 Idriess 1958, p 4.
13 Idriess 1958, p 5.
14 Idriess 1958, p 69.
15 Idriess 1958, p 70.
16 Idriess 1958, pp 115-116.
17 Idriess 1958, p 73.
18 Idriess 1958, p 75.
19 Idriess 1958, p 97.

Chapter 19 - First Fight, November 1920

1 Idriess, I 1938, *Madman's Island*, Angus and Robertson, Sydney, p 124.
2 Idriess, I 1927, *Madman's Island*, Cornstalk, Sydney.

3 Idriess 1927, pp 8-9, 12-15.
4 Idriess 1927, p 29.
5 Idriess 1938, p 100.
6 Idriess 1927, p 30. Idriess 1938, p 100.
7 Eley, B 1995, *Ion Idriess*, Harper Collins, Sydney, p 51.
8 Idriess 1938, pp 100-101.
9 Idriess 1938, p 102.
10 Idriess 1938, p 105.
11 Idriess 1927, p 34.
12 Idriess 1927, p 105-109, 134.

Chapter 20 - Mary Watson

1 Falkiner, S Oldfield, A 2000, *Lizard Island – The Journey of Mary Watson*, Allen & Unwin, Crows Nest.
2 Robertson, J 1981, *Lizard Island*, Hutchinson, Richmond.
3 *Cairns Post*, 2008? *The Heroine of Lizard Island*, *Cairns Post* Print.
4 Idriess, I 1938, *Madman's Island*, Angus and Robertson, Sydney, p 134. Idriess, I 1948, *The Opium Smugglers*, Angus and Robertson, Sydney, pp 138, 182.Idriess, I 1959, *The Tin Scratchers*, Angus and Robertson, Sydney, p 32. Idriess, I 1955, *The Vanished People*, Angus and Robertson, Sydney, p 1.
5 Falkiner & Oldfield 2000, p 52.
6 Idriess, I. 1957, *Coral Sea Calling*, Angus and Robertson, Sydney, pp 29, 50-53.
7 Falkiner & Oldfield 2000, p 60.
8 Falkiner & Oldfield 2000, p 63.
9 Falkiner & Oldfield 2000, p…
10 Robertson J 1981.
11 Robertson J 1981, p 123. Falkiner & Oldfield 2000, p 69.
12 Robertson J 1981, p 133.
13 Falkiner & Oldfield 2000, p 69.
14 Falkiner & Oldfield 2000, p 70.
15 Falkiner & Oldfield 2000, p 71.
16 Falkiner & Oldfield 2000, p 71.
17 Falkiner & Oldfield 2000, p 80.
18 Falkiner & Oldfield 2000, p 72.
19 Falkiner & Oldfield 2000, p 72.
20 Falkiner & Oldfield 2000, p 72.
21 Idriess, I 1948, *The Opium Smugglers*, Angus and Robertson, Sydney, p 139.
22 Robertson 1981, pp 132, 134.
23 National Trust of Queensland. *Posters*. Captain Cook Museum, Cooktown.
24 Falkiner & Oldfield 2000, pp 90-91.
25 Falkiner & Oldfield 2000, pp 85, 87.
26 Lucas, A. 1968, *Cruising the Coral Coast*, Horwitz, Cameray, p 291.
27 Falkiner & Oldfield 2000, pp 92-93.
28 Falkiner & Oldfield 2000, p 93.

29 Falkiner & Oldfield 2000, p 92.
30 Falkiner & Oldfield 2000, p 92.
31 Idriess 1955, p 6.
32 Falkiner & Oldfield 2000, p 92.
33 Falkiner & Oldfield 2000, p 92.
34 Falkiner & Oldfield 2000, p 92.
35 Falkiner & Oldfield 2000, p 93.
36 Falkiner & Oldfield 2000, p 93.
37 Falkiner & Oldfield 2000, p 93.
38 Falkiner & Oldfield 2000, p 93.
39 Falkiner & Oldfield 2000, p 93.
40 Falkiner & Oldfield 2000, p 98.
41 Falkiner & Oldfield 2000, p 74.
42 Falkiner & Oldfield 2000, p 75.
43 Falkiner & Oldfield 2000, p 75.
44 Falkiner & Oldfield 2000, p 75.
45 Falkiner & Oldfield 2000, pp 75-76.
46 Falkiner & Oldfield 2000, p 80.
47 Falkiner & Oldfield 2000, pp 74-81.
48 Falkiner & Oldfield 2000, p 88.
49 Falkiner & Oldfield 2000, p 98.
50 Falkiner & Oldfield 2000, p 100.
51 Falkiner & Oldfield 2000, p 101.
52 Falkiner & Oldfield 2000, p 104.

Chapter 21 - Wind

1 Idriess, I 1927, *Madman's Island*, Cornstalk, Sydney, p 33. Idriess, I 1938, *Madman's Island*, Angus and Robertson, Sydney, p 104.
2 Idriess 1927, p 34. Idriess 1938, p 108.

Chapter 22 - Cooktown

1 Idriess, I 1938, *Madman's Island*, Angus and Robertson, Sydney, p 108.
2 Idriess 1938, p 104.
3 Australian Electoral Rolls, 1901-1936, *George Tritton*, Ancestry.com.au.
4 *Cairns Post*, 2 May 1913 and 1 October 1914.
5 See Chapter Four.
6 Defence Service Records, *National Archives of Australia*, Tritton, George.
7 Idriess 1938, p 3.
8 Shay, J. & B. undated, *Pubs and publicans of Cooktown*, Cooktown and District Historical Society, pp 4, 19.
9 Shay J & B undated, p 28.
10 Idriess 1938, p 4.
11 Ryle, P, 2000, *Decline and Recovery of a Rural Coastal Town – Cooktown*, James Cook University, Townsville, p 382.
12 Idriess 1959, *The Tin Scratchers*, Angus and Robertson, Sydney, pp 2-5.

13 Idriess 1959, p 3.
14 Eley, B 1995, *Ion Idriess*, Harper Collins, Sydney, pp 67-69.
15 Idriess, I. 1963, *Our Living Stone Age*, Angus and Robertson, Sydney, p 59. Eley, B 1995, *Ion Idriess*, Harper Collins, Sydney, p 100.
16 Eley 1995, pp 69-70.
17 Idriess 1959, pp 10-13.
18 Defence Service Records, *National Archives of Australia*, Tritton, George. Queensland Registry, 1928, Births Deaths and Marriages, *George Trittan* sic, 287, 4229.
19 Idriess 1959, p 20.
20 Eley 1995, p 46.
21 Dick, A J, 2003, *Peninsula Pioneer*, Self-published, Jindalee, p 128.
22 Idriess, I 1962, *My Mate Dick*, Angus and Robertson, Sydney, pp 161, 199.
23 Dick 2003.
24 Dick 2003, p 118.
25 Dick 2003, p 119.
26 Dick 2003, p 120.
27 Dick 2003, pp 119-120.
28 Idriess 1959, p 21.
29 Idriess 1959, pp 21-30.
30 Idriess 1959, p 35.
31 Eley 1995, pp 108-109.
32 Idriess 1959, p 68.
33 Idriess 1959, p 68.
34 Idriess 1959, pp 69, 70, 84.
35 Idriess 1959, p 71.
36 Idriess 1959, p 86.
37 Idriess 1959, pp 92-95.
38 Idriess 1959, p 79.
39 Idriess 1959, p 104.
40 Idriess 1959, p 45.
41 Idriess 1959, pp 46, 127.
42 Idriess 1959, p 129.
43 Idriess 1959, pp 47, 135.
44 Idriess 1959, p 136.
45 http://www.yps.net.au/normanbaird/extracts.htm
46 Idriess 1959, p 260.
47 Eley 1995, p 69.
48 Eley 1995, p 71.
49 Eley 1995, p 73.
50 Eley 1995, pp 73-74.
51 Idriess 1938, p 109.

Chapter 23 - Second Fight, December 1920

1 Idriess, I 1938, *Madman's Island*, Angus and Robertson, Sydney, p 109.

2 Idriess, I 1927, *Madman's Island*, Angus and Robertson, Sydney, p 35. Idriess 1938, p 109.
3 Idriess 1927, p 35. Idriess 1938, p 110.
4 Idriess 1927, p 35. Idriess 1938, p 110.
5 Idriess 1927, p 36. Idriess 1938, p 113.
6 Idriess 1927, p 37.
7 Idriess 1938, p 113.
8 Idriess 1927, p 37. Idriess 1938, p 113.
9 Idriess 1938, p 113.
10 Eley, B 1995, *Ion Idriess*, Harper Collins, Sydney, pp 96-97. Growden, G 2003, *The Snowy Baker Story*, Random House, Milsons Point, p267.
11 Idriess 1927, p 37. Idriess 1938, p 113.
12 Idriess 1927, p 37. Idriess 1938, p 113.
13 *Cairns Post* 1921, Friday 18 February 1921.
14 Idriess 1938, p 2.
15 Idriess 1938, p 31.
16 Idriess 1938, p 100.
17 Idriess 1938, pp 99-100.

Chapter 24 - First Raid and Retaliation, December 1920

1 Idriess, I 1927, *Madman's Island*, Angus and Robertson, Sydney, pp 40-42. Idriess, I 1938, *Madman's Island*, Angus and Robertson, Sydney, pp 118-119.
2 Idriess 1938, pp 121-122.
3 Idriess 1938, p 125. See Chapter Five.
4 Idriess 1938, p 126.
5 Idriess 1938, p 128.
6 Idriess 1927, p 47.
7 Idriess 1938, p 128.
8 Idriess 1927, p 47.
9 Idriess 1927, p 47.
10 Idriess 1927, p 48.
11 Idriess 1927, p 49. Idriess 1938, pp 132-133.

Chapter 25 - Moynahan Sailed Away, December 1920

1 Idriess, I 1938, *Madman's Island*, Angus and Robertson, Sydney, p 10.
2 Idriess 1938, p 43.
3 Idriess 1938, pp 51-53
4 Idriess 1938, pp 146-147, 191-192.
5 Idriess 1938, p 191.
6 Idriess 1938, pp 139-144.
7 Idriess, I 1927, *Madman's Island*, Angus and Robertson, Sydney, pp 51-55. Idriess 1938, pp 140-141.
8 Idriess 1938, pp 140-141
9 Idriess 1927, pp 51-56.
10 Idriess 1938, p 141.

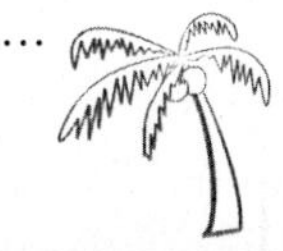

11 Idriess 1938, pp 142-144.
12 Idriess 1927, p 56. Idriess 1938, p 147.

Chapter 26 – Second Raid and Retaliation, December 1920

1 Idriess, I 1927, *Madman's Island*, Angus and Robertson, Sydney, p 58.
2 Idriess, I 1938, *Madman's Island*, Angus and Robertson, Sydney, p 150.
3 Idriess 1927, p 57.
4 Idriess 1938, pp 141-143.
5 Idriess 1927, p 57. Idriess 1938, p 149.
6 Idriess 1927, p 57. Idriess 1938, p 150.
7 Idriess 1938, p 151.
8 Idriess 1927, p 58.
9 Idriess 1938, p 151.
10 Idriess 1938, p 156.
11 Idriess 1927, pp 60-61. Idriess 1938, p 155.
12 Idriess 1927, p 62. Idriess 1938, p 156.
13 Idriess 1927, p 65. Idriess 1938, pp 159-162.
14 Idriess 1927, p 69. Idriess 1938, p 167.

Chapter 27 – Third Raid, January 1921

1 Idriess, I 1938, *Madman's Island*, Angus and Robertson, Sydney, p 168.
2 Idriess, I 1927, *Madman's Island*, Angus and Robertson, Sydney, p 72.
3 Idriess 1938, pp 51, 133, 147.
4 Idriess 1938, pp 10, 168.
5 Idriess 1927, p 72.
6 Idriess 1938, pp 169-173.
7 Idriess 1938, pp 182.
8 Idriess 1938, pp 173.
9 *Cairns Post* 1921, Friday 18 February 1921. National Library of Australia.
10 Defence Service Records, *National Archives of Australia*, Tritton, George.

Chapter 28 – War

1 Defence Service Records, Tritton, George.
2 http://www.awm.gov.au/encyclopedia/enlistment.
3 http://www.awm.gov.au/units/unit_11228.asp.
4 Defence Service Records, Tritton, George.
5 Defence Service Records, Tritton, George.
6 Defence Service Records, Tritton, George.
7 Defence Service Records, *National Archives of Australia*, Idriess, Ion Llewellyn.
8 Eley, B 1995, *Ion Idriess*, Harper Collins, Sydney, p 77
9 Idriess, I 1932a, *The Desert Column*, Angus and Robertson, Sydney, pp 272
10 Idriess 1932a, p 321.
11 Eley 1995, p 77-78
12 Idriess 1932a, p 173.

13 Idriess 1932a, p 122.
14 Idriess, I. 1934, *The Yellow Joss*, Angus and Robertson, Sydney, p 18.
15 Eley 1995, p 75.
16 Idriess 1932a, p 3.
17 Idriess 1932a, p 13.
18 Idriess 1932a, p 17.
19 Idriess 1932a, p 18.
20 Idriess 1932a, p 21.
21 Idriess 1932a, p 18.
22 Idriess 1932a, p 23.
23 Idriess 1932a, p 25.
24 Idriess 1932a, pp 27-39.
25 Idriess 1932a, p 39.
26 Hamilton, J 2008, *Gallipoli Sniper*, Macmillan, Sydney, p 215.
27 Idriess 1932a, p 137.
28 Idriess 1932a, pp 51-55.
29 Jeffery, M, 2006, *Address on the occasion of a Memorial Service*, Lone Pine, Turkey, 25 April 2006.
30 Idriess 1932a, p 56.
31 Idriess 1932a, p 61.
32 Idriess 1932a, p 66.
33 Idriess 1932a, p 68.
34 Idriess 1932a, p 69.
35 Idriess 1932a, p 105.
36 Idriess 1932a, p 121.
37 Idriess 1932a, p 122.
38 Idriess 1932a, p 123.
39 Idriess 1932a, p 133.
40 Idriess 1932a, p 135.
41 Idriess 1932a, p 139.
42 Idriess 1932a, p 202.
43 Idriess 1932a, p 232.
44 Idriess 1932a, p 251.
45 Idriess 1932a, p 252.
46 Idriess 1932a, p 252.
47 Idriess 1932a, p 288.
48 Idriess 1932a, p 272.
49 Idriess 1932a, p 321.
50 Idriess 1932a, p 325.
51 Idriess 1932a, p 325.
52 http://alh-research.tripod.com/Light_Horse/index.blog/1816069/ion-idriess-and-the-beersheba-charge-description/.http://desert-column.phpbb3now.com/viewtopic.php?f=2&t=479
53 Australian War Memorial Research Centre, IDRL/0373.
54 Idriess 1932a, p 15. http://www.spirits-of-gallipoli.com

55 Idriess 1932a, pp 331-332.
56 Idriess 1932a, p 339.
57 Idriess 1932a, p 345.
58 Idriess 1932a, pp 375-379. Eley 1995, p 96.

Chapter 29 - Mud Pool, January 1921

1 Idriess 1938, p 171-172.
2 Idriess 1927, p 73. Idriess 1938, p 186.
3 Idriess, I 1938, *Madman's Island*, Angus and Robertson, Sydney, p 187.
4 Idriess 1938, p 189.
5 Idriess, I 1927, *Madman's Island*, Angus and Robertson, Sydney, p 71.
6 Idriess 1938, p 178.
7 Idriess 1938, pp 131-132.
8 Idriess 1927, p 75.
9 Idriess 1927, p 75. Idriess 1938, p 189.
10 Personal communication with Peter Rutherford. See acknowledgements.
11 Idriess 1938, p 182.
12 Idriess 1938, p 174.
13 Idriess 1927, p 74.
14 Personal communication with Peter Rutherford. See acknowledgements.
15 Idriess 1927, p 75. Idriess 1938, p 188-189.
16 Idriess 1927, p 76. Idriess 1938, p 189.
17 Idriess 1927, p 77.
18 Idriess 1927, pp 75-77.
19 Idriess 1938, pp 188.
20 Idriess 1938, pp 189.
21 Idriess 1927, p 77.
22 Idriess 1938, p 190.
23 Idriess 1927, p 78.

Chapter 30 - After the War

1 Defence Service Records, *National Archives of Australia*, Tritton, George.
2 *Cairns Post*, 11 October 1919, p 4.
3 *Cairns Post*, 31 October 1919, p 6.
4 http://www.aif.adfa.edu.au.
5 Australian Death Index, *William Henry Tritton*, 000598, Ancestry.com.au.
6 Australian Electoral Rolls, 1901-1936, *George Tritton*, Ancestry.com.au.
7 Eley, B 1995, *Ion Idriess*, Harper Collins, Sydney, p 96.
8 Eley 1995, pp 98,100.
9 Growden, G. 2003, *The Snowy Baker Story*, Random House, Milsons Point, p 267.
10 Eley 1995, p 97.
11 Eley 1995, p 98.
12 *Brisbane Courier*, 18 October 1919, p 3.
13 *Cairns Post*, 16 March 1920, p 4.

14 Eley 1995, p 98.
15 Eley 1995, p 99.
16 Personal communication with Jim McJannett. See acknowledgements.
17 http://www.aiatsis.gov.au.
18 Yindillin Professional Services, http://www.yps.net.au/normanbaird/extracts.htm.
19 Pearson, N 2006, "Walking in two Worlds", *The Australian*, October 28, 2006.
20 Pearson 2006.

Chapter 31 - Opium, January 1921

1 Idriess, I 1934, *The Yellow Joss*, Angus and Robertson, Sydney. Idriess, I 1948, *The Opium Smugglers*, Angus and Robertson, Sydney.
2 Idriess, I 1927, *Madman's Island*, Angus and Robertson, Sydney, pp 80-81, 84-85.
3 Idriess 1927, pp 85-87, 93-94.
4 Idriess 1927, pp 111-112, 129-133, 158-159, 180-207.
5 Idriess 1927, p 227.
6 Idriess, I 1938, *Madman's Island*, Angus and Robertson, Sydney, pp 197-198.
7 Idriess 1938, p 210.
8 Idriess 1927, p 83.
9 Idriess 1938, p 199-201.
10 Idriess 1938, p 210.
11 Idriess 1938, pp 206-210.
12 Idriess 1938, p 211.
13 Idriess 1938, p 211-219.

Chapter 32 - Thoughts and Feelings

1 Idriess, I 1927, *Madman's Island*, Angus and Robertson, Sydney, p 12.
2 Idriess 1927, p 13.
3 Idriess 1927, p 15.
4 Idriess 1927, p 90.
5 *Cairns Post* 1921, Friday 18 February 1921.
6 Idriess, I 1938, *Madman's Island*, Angus and Robertson, Sydney, p 10.
7 Idriess 1938, p 54.
8 Idriess 1938, pp 141-143.
9 Idriess 1938, p 213.
10 Idriess 1938, pp 208, 215.
11 Idriess 1938, p 214.
12 Idriess 1938, p 216.
13 Idriess 1938, p 211.
14 Eley, B 1995, *Ion Idriess*, Harper Collins, Sydney, p 161.
15 Eley 1995, p 160.
16 Eley 1995, p 237.
17 Idriess, I 1933, *Drums of Mer*, Angus and Robertson, Sydney, pp 146-150.
18 Idriess, I 1932a, *The Desert Column*, Angus and Robertson, Sydney, p 211.

19 Idriess 1932a, pp 213-214.
20 Eley 1995, pp 191-192.
21 Robertson, I, 1991, *What the Cults Believe*, Moody Publishers, p 179.
22 Eley 1995, p 367.
23 Eley 1995, p 367.
24 Eley 1995, p 161.
25 Eley 1995, p 161.
26 Eley 1995, p 367.
27 Idriess 1938, p 219.
28 Idriess 1938, p 219.

Chapter 33 – Rescue, February 1921

1 Idriess, I 1938, *Madman's Island*, Angus and Robertson, Sydney, p 211..
2 Idriess, I 1927, *Madman's Island*, Angus and Robertson, Sydney, pp 90-91.
3 Idriess 1938, p 237.
4 Idriess 1938, pp 233, 236, 239.
5 Idriess 1938, pp 241-242

Chapter 34 – George's "Madness"

1 Idriess, I 1938, *Madman's Island*, Angus and Robertson, Sydney, p 217.
2 Idriess 1938, p 218.
3 The *Brisbane Courier* (Qld), Monday 28 February 1921.
4 Personal communication with Jim McJannett. See acknowledgements.
5 *Cairns Post* (QLD), 2 May 1922, p 8. *Cairns Post* (QLD), 2 September 1925, p 12

Chapter 35 – Jack's Emotional State

1 Diagnostic and Statistical Manual of Mental Disorders, Fourth Edition DSM-IV
2 www.medicinenet.com/posttraumatic_stress_disorder/article.htm.
3 psychology.jrank.org/pages/499/Post-Traumatic-Stress-Disorder-PTSD.html>
4 *American Journal of Psychiatry*, 1995, Risk Factors for PTSD-related Traumatic Events: a prospective analysis. N Breslau, N et al, 152:529-535. *Journal of Traumatic Stress*, 1991, Predisposing Variables in PTSD Patients, Emery, V et al, Springer Netherlands. Vol.4, No. 3.
5 Eley, B 1995, *Ion Idriess*, Harper Collins, Sydney, p 81.
6 Hetherington, J 1963, *Forty-two Faces*, Angus and Robertson, London, p 13.
7 Idriess, I. 1932a, *The Desert Column*, Angus and Robertson, Sydney.
8 Eley 1995, e.g. Chapters 32-35.

Chapter 36 – What Happened Next?

1 The *Brisbane Courier* (Qld), Monday 28 February 1921.
2 *Cairns Post* (QLD), 20 October 1921, p 4.
3 *Cairns Post* (QLD), 20 October 1921, p 4.
4 *Cairns Post* (QLD), 2 September 1925, p 12.

5 Australian Electoral Rolls, 1901-1936, *George Tritton*, Ancestry.com.au.
6 Australian Electoral Rolls, 1901-1936, *George Tritton*, Ancestry.com.au.
7 Eley, B 1995, *Ion Idriess*, Harper Collins, Sydney, pp 106-111.
8 Eley 1995, pp 102-105.
9 Eley 1995, pp 102.
10 Eley 1995, pp 102-105.
11 *Cairns Post* (QLD), 24 June 1921, p 8.
12 The Sydney Morning Herald (NSW), 29 March 1923, p 11.
13 The *Brisbane Courier* (Qld), 31 March 1923 p 7.
14 Bradly, J ed. 2008, *Gouger of the Bulletin*, Idriess Enterprises, Uralla, p 69.
15 Darroch, R 1981, *D. H. Lawrence in Australia*, Macmillan, Sth. Melb, pp 68-70.
16 Bradly 2008, p 66.
17 Barker, A 1982, *George Robertson*, UQ Press, Brisbane, p 169.
18 Bradly 2008, pp 82-83.
19 Eley 1995, p 105.
20 Eley 1995, p 111.
21 Eley 1995, pp 116-118.
22 Eley 1995, p 118.
23 Eley 1995, p 119.
24 Eley 1995, p 131.
25 Eley 1995, p 131.
26 Eley 1995, pp 324-325.
26 Eley 1995, p 133.
27 Barker 1982, p 169.
28 Bradly 2008.
29 Eley 1995, pp 102-104.
30 Barker 1982, p 170.
31 Eley 1995, p 102.
32 Eley 1995, p 104.
33 Eley 1995, p 369.
34 Community Research Unit, 2005, *Frankston's Favourite Book*, City of Frankston.
35 For Example: Gleeson-White J, (2007), *Australian Classics*, Allen and Unwin, Sydney. Laird J (ed), (1988), *The Australian Experience of War*, Mead and Beckett Publishing, Sydney. Lyons M and Arnold J, *A History of the Book in Australia*, University of Queensland Press Brisbane, 2001. University of Sydney, 2005, *Classic Australian Works*, Sydney University Press, AUS. Webby E, 2000, *Australian Literature*, Cambridge University Press, London. Williamson, G, 2012, *The Burning Library*, Text Publishing, Melbourne.
36 Barker 1982, pp 186-187.
37 Eley 1995, p 148.
38 Burnet, R, 1996, *An Idriess Bibliography*, Australian Book Collector, Uralla.
39 Carlyle, T, 1888, *On Heroes, Hero-worship and the Heroic in History*, Frederick A Stokes and Brother, New York.

40 Carlyle 1888, p 177.

Chapter 37 - The Truth about Charlie

1 Idriess, I 1938, *Madman's Island*, Angus and Robertson, Sydney, p VII.
2 Personal communication with Jim McJannett – see Acknowledgements.
3 The *Brisbane Courier* (Qld), Monday 16 February 1921, p 6
4 Eley, B 1995, *Ion Idriess*, Harper Collins, Sydney, pp 100-102. Bowden, T 1972? *http://www.abc.net.au/rn/radioeye/documents/idriess.pdf*
5 *Cairns Post* (Qld) Friday 18 February 1921, p 4.
6 The *Brisbane Courier* (Qld), Monday 28 February 1921, p 6.
7 Eley 1995, p 100.
8 Eley 1995, p 303.
9 Eley 1995, p 130.
10 Personal communication with Peter Rutherford – see Acknowledgements.
11 Clemens, S. L. (Mark Twain) (1885), *The Adventures of Huckleberry Finn*, Charles L Webster and Company, NY.
12 See Chapter Six

About the author

The author, Rob Coutts, lives in Brisbane. He is married, with two children living in nearby suburbs and two grand-daughters. Professionally, Rob has his own social work practice providing counselling assistance to war veterans and war widows.

His interest in Ion Idriess began sixty-four years ago when he was a paper-boy delivering newspapers in the Melbourne suburb of Ashburton. For his ninth birthday one of his customers gave him Idriess' *In Crocodile Land.* It was the imagery of "The Blood Hole" (where crocodiles waited at the gruesome blood drain of the meat works at Wyndham) that began a lifelong interest in Idriess' books. Rob still has that book as part of his collection.

Rob administers the website: *idriess.com.au*